THE LIBRARY
ST. MARY'S COLLEGE OF MARYLAND
ST. MARY'S CITY, MARYLAND 20686

Northwest Atlantic Groundfish: Perspectives on a Fishery Collapse

Northwest Atlantic Groundfish: Perspectives on a Fishery Collapse

Edited by

John Boreman, Brian S. Nakashima,
James A. Wilson, and Robert L. Kendall

American Fisheries Society
Bethesda, Maryland

1997

Suggested citation formats follow.

Entire book

Boreman, J., B. S. Nakashima, J. A. Wilson, and R. L. Kendall, editors. 1997. Northwest Atlantic groundfish: perspectives on a fishery collapse. American Fisheries Society, Bethesda, Maryland.

Chapter within the book

Drinkwater, K. F., and D. G. Mountain. 1997. Climate and oceanography. Pages 3–25 in J. Boreman, B. S. Nakashima, J. A. Wilson, and R. L. Kendall, editors. Northwest Atlantic groundfish: perspectives on a fishery collapse. American Fisheries Society, Bethesda, Maryland.

Cover photograph: Boston Fish Market, 1935. From the Oscar E. Sette Collection, courtesy of the National Marine Fisheries Service, Northeast Fisheries Science Center, Woods Hole, Massachusetts.

© Copyright by the American Fisheries Society, 1997

All rights reserved. Photocopying for internal or personal use, or for the internal or personal use of specific clients, is permitted by AFS provided that the appropriate fee is paid directly to Copyright Clearance Center (CCC), 222 Rosewood Drive, Danvers, Massachusetts 01923, USA; phone 508-750-8400. Request authorization to make multiple photocopies for classroom use from CCC. These permissions do not extend to electronic distribution or long-term storage of articles or to copying for resale, promotion, advertising, general distribution, or creation of new collective works. For such uses, permission or license must be obtained from AFS.

Library of Congress Catalog Card Number: 97-77481
ISBN 1-888569-06-9
Printed in the United States of America on recycled, acid-free paper

Published by the American Fisheries Society

CONTENTS

Preface vii
Acknowledgments ix
Chapter Summaries xi
Contributors and Editors xvii
Book Conventions xix
Species List xxi

PART I THE ONCE AND PRESENT FISHERY

1. Climate and Oceanography 3
   K. F. Drinkwater and D. G. Mountain

2. Groundfish Stocks and the Fishing Industry 27
   S. A. Murawski, J.-J. Maguire, R. K. Mayo, and F. M. Serchuk

3. Why Have Groundfish Stocks Declined? 71
   A. F. Sinclair and S. A. Murawski

4. Policy Frameworks 95
   R. G. Halliday and A. T. Pinhorn

5. Role of Local Institutions in Groundfish Policy 111
   M. Hall-Arber and A. C. Finlayson

PART II MANAGEMENT CONSIDERATIONS FOR REHABILITATED STOCKS

6. Single-Species and Multispecies Management 141
   B. J. Rothschild, A. F. Sharov, and M. Lambert

7. Ecosystem-Based Management 153
   R. W. Langton and R. L. Haedrich

8. Fishery Reserves 159
   P. J. Auster and N. L. Shackell

9	Fishing Gear Management	167
	J. T. DeAlteris and D. L. Morse	
10	User Rights in Fishing	177
	R. E. Townsend and A. T. Charles	
11	Comanagement	185
	L. Felt, B. Neis, and B. McCay	
12	Corporate Management of Fisheries	195
	R. E. Townsend	
13	Decision Analysis	203
	D. E. Lane and R. L. Stephenson	

Bibliography	211
Glossary	233
Index	237

PREFACE

The drive toward perpetual expansion—or personal freedom—is basic to the human spirit. But to sustain it we need the most delicate, knowing stewardship of the living world that can be devised.

Edward O. Wilson (1984)

Fishery collapse, closure, defaults on loans, vessel buyouts, and unemployment—a scenario that is becoming all too familiar to fishing communities worldwide. For the groundfish fisheries of the northwest Atlantic, the scenario has reached crisis proportions. Landings of principal groundfish are now at record low levels in the United States and Canada for the majority of the traditionally important groundfish species. As a result, continuation of a way of life that has existed in fishing communities from Labrador to New England for 300–400 years is being threatened. The capability to regulate the fisheries effectively is being outpaced by the capability to find groundfish and catch them, as fishery managers struggle to keep pace with technology and market demand. Management agencies in the United States and Canada are now laying paths out of the groundfish crisis, but all stakeholders in the resource need to be looking for a means to prevent a recurrence.

Who are the stakeholders? The status of the groundfish resource is important to a variety of sectors in our society. Within the industry sector, it affects those people who catch, process, sell, and consume the fish, and the communities in which those people live. The government sector is concerned with conservation and sustainable utilization of the resource and with the balance between the resource's abundance and the status of its predators, competitors, and prey in the northwest Atlantic ecosystem. To nongovernment organizations, the groundfish species represent fundamental components of the biodiversity of nature that has taken a billion years to evolve. Essentially, the well-being of the groundfish resource affects us all.

The American Fisheries Society (AFS), the major association of fishery professionals in North America, is committed to promoting the conservation, development, and wise use of fisheries resources by producing information in accessible forms and fostering dialogue and partnerships. At the 124th annual meeting of AFS, held during August 1994 in Halifax, Nova Scotia, fisheries scientists, resource managers, oceanographers, economists, anthropologists, and sociologists began an intensive and coordinated examination of the groundfish crisis in the northwest Atlantic. Immediately apparent to those present at the meeting was that any resolution will require cooperation among all the stakeholders, which will include sharing of information, creating open and willing dialogues, and developing common, acceptable, and achievable goals.

Subsequent to the Halifax meeting, AFS formed a groundfish steering committee representing those of its subunits with the greatest professional interest in groundfish: the Canadian Aquatic Resources Section, the Marine Fisheries Section, the Socioeconomics Section, the Northeastern Division, the Atlantic International Chapter, and the Southern New England Chapter. This book fulfills the committee's charge to provide an informational resource for public discussions about rehabilitating and sustaining the now-depleted groundfish stocks of the northwest Atlantic. Addressing both Canada and the United States, the book describes the climatic and oceanographic conditions in the region prior to and during the collapse of the fisheries. It covers the history of the fisheries, their human dimensions, the regulatory and policy frameworks in which they are embedded, and management options that may be considered for their future. The book discusses, as factually as possible, how and why the groundfish crisis developed and what the impact of the crisis has been on the groundfish resources, the communities that rely on those resources, and the economic stability of the northwest Atlantic region. The book is intended to describe the problems, not dictate the answers, to an audience that includes interested citizens who lack training in the biological, social, and political sciences. The authors of each chapter have been instructed to distill the knowledge relevant to the issue and treat matters dispassionately, fairly, and without aspersion.

The committee advocates a consensus-building approach among all the stakeholders to the future conservation and utilization of the groundfish resource. Perhaps the current system of management will work with some adjustment. Perhaps new systems should be explored. We hope this book will lead to a common understanding of the issues and that it will stimulate discussion and agreement on workable solutions, resulting in a step toward Wilson's "most delicate, knowing stewardship."

JOHN BOREMAN
BRIAN S. NAKASHIMA
JAMES A. WILSON
ROBERT L. KENDALL

ACKNOWLEDGMENTS

Creation of this book has been a binational endeavor of American Fisheries Society (AFS) members and their colleagues. A project of the Society's Fishery Information Network, the book has been strongly supported by the AFS Marine Fisheries Section, Canadian Aquatic Resources Section, Socioeconomics Section, Northeastern Division, Atlantic International Chapter, and Southern New England Chapter. A general steering committee guided the planning and development of the project. Under the chairmanship of John Boreman, this committee was served by Nancy Adams, Richard Allen, Dale Burkett, Scott Decker, Peter Fricke, Graeme Gawn, William Hyatt, Brian Nakashima, Howard Powles, and James Wilson; John Fritts, Robert Kendall, and Pamela McClelland provided AFS staff support. An editorial subcommittee consisting of Boreman, Nakashima, Powles, Wilson, and Kendall worked with authors and reviewers in the crafting of chapters.

All the chapters were peer reviewed for technical accuracy and clarity. Frank Aikman, Spencer Apollonio, Bradley W. Barr, John Gates, Stratis Gavaris, Philip E. Haring, Ray Hilborn, Jeffrey A. Hutchings, John F. Kearney, Bruce M. Leaman, W. H. Lear, Gordon Mertz, Bruce Rettig, Jake Rice, Michael Sinclair, Grant G. Thompson, and George H. Winters all contributed helpful and constructive comments that led to substantial improvements in the book. Editorial committee members and some authors also provided reviews of various chapters. Carolyn Creed, Lloyd M. Dickie, Peter Fricke, and Stephen R. Kerr made strong intellectual contributions to the book during its early development. By all rights, Howard Powles's name should be on the cover of this volume. He has declined this formal credit, but he remains a coeditor in the minds of his colleagues.

The AFS Fishery Information Network and its "Stocks at Risk" program were under the direction of Pamela McClelland during the early and maturing stages of the groundfish project, and she did much to keep the project on track. John Fritts, with the help of McClelland and AFS Executive Director Paul Brouha, secured funding for the book. Janet Harry ably composed and corrected the several drafts of this book, and Beth Staehle guided its design and publication.

Important underwriting for the program and for publication of the book was provided by the

National Marine Fisheries Service
Surdna Foundation
National Fish and Wildlife Foundation
Town Creek Foundation
Norcross Wildlife Foundation

Without this financial help—and the strong efforts of authors, steering committee members, reviewers, and others named above—this book would have remained simply an idea.

CHAPTER SUMMARIES

CHAPTER 1 CLIMATE AND OCEANOGRAPHY

Changes in climate, particularly in temperature, affect the distribution, growth, and recruitment of northwest Atlantic groundfish. Air and sea temperatures off Labrador and northern Newfoundland generally have cooled since the 1960s, showing progressively lower minima in the early 1970s, the mid-1980s, and the 1990s. The general temperature decline and the minima have been accompanied by southward extensions of sea ice, and they are associated with stronger northwest winds over the Labrador Sea and with intensification of large-scale atmospheric circulation over the North Atlantic Ocean. The Labrador Current carries cold subarctic water southward over the continental shelf and slope from Hudson Strait to Cape Hatteras. Along the way, some slope water is exchanged with oceanic water, especially (south of the Grand Bank) with water transported northward by the warm Gulf Stream; slope water also is exchanged with shelf water. These exchanges account for much of the year-to-year variability in water characteristics on the continental shelf. The magnitude of water exchange is influenced by the strength of the Labrador Current, and this strength varies with subarctic atmospheric conditions. Because of the Labrador Current's southward flow, groundfish stocks throughout the northwest Atlantic region are affected by climatic and oceanographic events to their north.

CHAPTER 2 GROUNDFISH STOCKS AND THE FISHING INDUSTRY

Groundfish resources of the northwest Atlantic are severely depleted and may take a decade to recover. For more than 300 years, Canadian and U.S. fisheries exploited Atlantic cod with baited hooks. With the introduction of bottom trawls in the Twentieth Century, catches of other species increased and new markets and processing methods were developed. Beginning in the late 1950s, distant-water fleets from other countries increasingly exploited stocks beyond U.S. and Canadian territorial seas, which then extended less than 22 km from shore. Landings of "principal" groundfish (chiefly cod, haddock, and redfish) peaked in 1965 at 2.1 million tons and flounder landings peaked in 1968 at 315,000 tons; then landings of both groups declined precipitously. After Canada and the United States extended their exclusive marine jurisdictions out to 370 km in 1977, stocks recovered for a few years but then declined again as domestic fishing effort rapidly developed. In 1993, landings dropped to 16% of the historical maximum for principal groundfish and to 35% for flounders. Fishing for many Canadian groundfish stocks has been severely reduced or halted. In the United States, some fishing areas have been closed, and other constraints on fishing effort have been implemented. Both countries have begun to permanently reduce fishing capacity. Though many fishers have switched to other species, the rate at which depleted groundfish stocks will recover remains uncertain.

Chapter 3 WHY HAVE GROUNDFISH STOCKS DECLINED?

The most important cause of spectacular recent declines in northwest Atlantic groundfish resources was a fishing intensity so great that the stocks could not, on average, reproduce themselves to their former abundances. Persistent high harvests of mature fish and wasteful captures and discarding of immature fish have reduced the stocks' abilities to overcome natural variability in recruitment. Environmental change likely affected stock production, especially in northern waters, and may be influencing stock recovery, but no environmental factor alone can explain either the general decline in groundfish productivity since the 1950s or the precipitous decline in the 1990s. Species interactions such as predation and competition were negligible contributors to these declines. Almost without exception, fishing mortality rates have exceeded sustainable levels. Early warning signs of stress on groundfish populations included truncated age structures, altered growth rates, earlier sexual maturation, and increased variability in catches as fisheries became ever more dependent on the strength of incoming year-classes. Stock rebuilding may require a decade or more of harvest rates at or very near zero.

Chapter 4 POLICY FRAMEWORKS

Northwest Atlantic groundfish stocks became sharply depleted in both Canadian and U.S. waters despite markedly different fisheries policies in the two countries. Both countries were members of the International Commission for the Northwest Atlantic Fisheries from 1950 through 1976. The Commission had jurisdiction beyond the 22-km territorial limits then extant, but it was unable to control groundfish exploitation by distant-water fleets. The United States and Canada extended their jurisdictions to 370 km in 1977, and a new body—the Northwest Atlantic Fisheries Organization (NAFO)—was created to manage international fisheries seaward of the new limits. Canada joined NAFO immediately; within national waters, it combined a low-exploitation fishery strategy with a high-employment social policy that required large-scale federal intervention in management. The United States did not join NAFO until late 1995; in its waters, it delegated groundfish management to regional, nonfederal councils, which regulated exploitation indirectly through gear regulations and area closures. The outcomes of both national policies has been unsatisfactory, and both countries are implementing changes. Meanwhile, NAFO has been largely ineffective in managing international harvests. Noncompliance with harvest regulations has undermined groundfish policy in all jurisdictions.

Chapter 5 ROLE OF LOCAL INSTITUTIONS IN GROUNDFISH POLICY

In both the United States and Canada, local fishers react to national groundfish policies according to local economic and social norms and histories. Studies of fishing communities in Massachusetts and Nova Scotia demonstrate how local institutions can support, undermine, or otherwise cope with externally imposed fisheries policies. In general, federal fisheries policies work best when fishers believe they have been fully consulted in the development of fishery rules and that they have a say in the implementation of those rules. If local authorities have the endorsement of local interests, they believe they can enforce the rules. A form of comanagement involving local fisheries interests and federal authorities may be a necessary condition for sustainable groundfish harvest management in the northwest Atlantic.

CHAPTER 6 SINGLE-SPECIES AND MULTISPECIES MANAGEMENT

The strong reliance of fishery managers on results from mathematical population models did not prevent the collapse of groundfish resources throughout the northwest Atlantic region. Groundfish have been managed as individual species, typically as individual stocks of those species, to the present day. Modern single-species management relies on increasingly sophisticated mathematical models to estimate the size of a stock and the quantity of fish that can be harvested from it, but these models cannot account for important interactions among species and fishery sectors or for any environmental effects that might exist. Development of multispecies models for some of Europe's North Sea fisheries has shown that species interactions (predator–prey relationships in particular) have a much stronger influence on fishery yields than had been appreciated. Multispecies models are becoming more "realistic," but they are still incomplete. They require very large amounts of data, which are expensive to collect, and their practicality remains questionable. However, some form of multispecies management—at least to the extent of optimizing joint harvests of key species—may be possible in the northwest Atlantic once groundfish stocks have recovered.

CHAPTER 7 ECOSYSTEM-BASED MANAGEMENT

The collapse of U.S. and Canadian groundfish stocks in the northwest Atlantic highlights the need for alternative management approaches. Ecosystem-based management will balance biological and environmental interactions with human social and economic values to maximize the sustainable yield of ecosystems. Hitherto, marine fisheries have been driven more by economic return and narrow biological considerations than by broad ecological principles. Requirements for successful ecosystem-based management include recognition of biological limits to fish production, clear statements of relevant human values, and a policy for governing human behavior.

CHAPTER 8 FISHERY RESERVES

Seasonally or permanently closed fishery reserves can enhance conservation of harvested species. Fishery reserves can sustain marine resources and their support systems, protect spawning and nursery areas of key species, maintain age structures (retain older, more fecund individuals), protect key habitats, and reduce bycatch. Reserves currently are used sporadically in the management of northwest Atlantic fisheries and are generally based on analyses of population data. However, important features of marine fish habitats are still poorly identified.

Fishers and scientists already have partial information on the life history and behavior of commercial species, on species assemblage patterns, and on physical structures of the marine environment. This information provides a starting point for identifying areas of particular importance to fish stocks or assemblages, leading to an experimental system of fishery reserves.

CHAPTER 9   FISHING GEAR MANAGEMENT

Groundfish management will be enhanced if fishing gear can be made more selective for particular species or fish sizes. Recent developments in fishing gear technology have made groundfish harvesting highly efficient, profoundly affecting the ocean system. Although they catch fish efficiently, many gears are not very selective in the species or sizes of fish they catch. The often high levels of bycatch and discarding that result from catches of mixed species and sizes greatly compromise management policies. Selectivity is encouraged by markets, which direct effort to certain species, and by fishery regulations, which constrain the times and places effort can be expended. Selectivity can be greatly increased, however, by gear modifications based on fish size and an understanding of each species' behavior in relation to the gears.

CHAPTER 10   USER RIGHTS IN FISHING

User rights can reduce the incentives for overfishing that occur in open-access fisheries. In many marine societies, social institutions regulated access to fishing long before Western concepts of fisheries management developed. The institutions of "customary marine tenure" and "territorial use rights in fishing" are usually community based and established by cultural norm. Governmental initiatives to create user rights through limited entry regulations became widespread in the 1960s. Limited entry has not fully controlled fishing pressure because each permit owner has an incentive to add more gear. Better-defined user rights include individual gear use rights (which limit some components of fishing effort for each fisher) and individual harvest quotas (including individual transferable quotas). These forms of user rights generate more economic benefits for fishers than simple limited entry, but they raise new concerns about high-grading and concentration of ownership.

CHAPTER 11   COMANAGEMENT

Sustainable fisheries management requires full cooperation and support from those who are being managed. Comanagement is any administrative system that gives resource users a meaningful role in the management of their resources. In comanagement regimes, users share with government both the power to make decisions and accountability for those decisions. Comanagement systems can enhance the scientific understanding of fisheries resources, the effectiveness of management initiatives, and the compliance with those initiatives. It also can deal with issues of social justice and equity that are not addressed by initiatives to privatize fisheries resources.

Chapter 12   CORPORATE RIGHTS

Fisheries could be governed by corporations. Such corporations would operate under a renewable long-term contract with government written to protect the public interest in ecological integrity. Corporate management replaces the resource-destructive incentives of open-access fisheries with an institution for making joint decisions and securing long-term rights. A corporate approach fundamentally shifts the government role from command-and-control regulator to architect of appropriate social institutions. Because appropriate incentives are created, a comprehensive set of responsibilities can be assigned to industry, including resource management strategies, effort or harvest controls, and allocation decisions. The corporation would also have responsibility for financing research, administration, and enforcement.

Chapter 13   DECISION ANALYSIS

Decision analysis, a field of study that treats complex problems, can assist fisheries management by providing a structured framework for the systematic resolution of problems. Better decisions should result. Use of this framework forces clear identification of fishery problems, explicit statement of fishery objectives, and evaluation of alternative strategic solutions of problems. Past decisions are judged by measuring their ability to achieve stated objectives. An effective decision process anticipates dynamic changes in the management system and monitors the system's ongoing progress toward specified objectives.

CONTRIBUTORS AND EDITORS

Peter J. Auster (Chapter 8): National Undersea Research Center for the North Atlantic and Great Lakes, The University of Connecticut at Avery Point, Groton, Connecticut 06340, USA

John Boreman (Coeditor): National Marine Fisheries Service, Northeast Fisheries Science Center, Woods Hole, Massachusetts 02543, USA

Anthony T. Charles (Chapter 10): Professor of Management Science, Department of Finance and Management Science, Saint Mary's University, Halifax, Nova Scotia B3H 3C3, Canada

Joseph T. DeAlteris (Chapter 9): Department of Fisheries, Animal and Veterinary Science, Fisheries Center, East Farm Campus, University of Rhode Island, Kingston, Rhode Island 02881, USA

Ken F. Drinkwater (Chapter 1): Department of Fisheries and Oceans, Bedford Institute of Oceanography, Box 1006, Dartmouth, Nova Scotia B2Y 4A2, Canada

Larry Felt (Chapter 11): Department of Sociology, Memorial University, St. John's, Newfoundland A1C 5S7, Canada

A. Christopher Finlayson (Chapter 5): Maine Department of Marine Resources, State House, Station 21, Augusta, Maine 04333-0021, USA

Richard L. Haedrich (Chapter 7): Department of Biology, Memorial University of Newfoundland, St. John's, Newfoundland A1B 3X7, Canada

Madeleine Hall-Arber (Chapter 5): Sea Grant College Program, Massachusetts Institute of Technology, 292 Main Street, E38-300, Cambridge, Massachusetts 02139, USA

Ralph G. Halliday (Chapter 4): Science Branch, Department of Fisheries and Oceans, Bedford Institute of Oceanography, Post Office Box 1006, Dartmouth, Nova Scotia B2Y 4A2, Canada

Robert L. Kendall (Coeditor): American Fisheries Society, 5410 Grosvenor Lane, Suite 110, Bethesda, Maryland 20814, USA

Marjorie Lambert (Chapter 6): University of Massachusetts - Dartmouth, 285 Old Westport Road, North Dartmouth, Massachusetts 02747-2300, USA

Daniel E. Lane (Chapter 13): Faculty of Administration, University of Ottawa, Ottawa, Ontario K1N 6N5, Canada

Richard W. Langton (Chapter 7): Maine Department of Marine Resources, Marine Resources Laboratory, Post Office Box 8, West Boothbay Harbor, Maine 04575, USA

Jean-Jacques Maguire (Chapter 2): 1450 Godefroy, Sillery, Quebec G1T 2E4, Canada

xviii   Contributors and Editors

Ralph K. Mayo (Chapter 2): National Marine Fisheries Service, Northeast Fisheries Science Center, Woods Hole, Massachusetts 02543, USA

Bonnie McCay (Chapter 11): Department of Human Ecology, Cook's College, Rutgers University, New Brunswick, New Jersey 08903, USA

Dana L. Morse (Chapter 9): Department of Fisheries, Animal and Veterinary Science, Fisheries Center, East Farm Campus, University of Rhode Island, Kingston, Rhode Island 02881, USA

David G. Mountain (Chapter 1): National Marine Fisheries Service, Northeast Fisheries Science Center, Woods Hole, Massachusetts 02543, USA

Steven A. Murawski (Chapters 2 and 3): National Marine Fisheries Service, Northeast Fisheries Science Center, Woods Hole, Massachusetts 02543, USA

Brian S. Nakashima (Coeditor): Department of Fisheries and Oceans, Northwest Atlantic Fisheries Centre, Post Office Box 5667, St. John's, Newfoundland A1C 5X1, Canada

Barbara Neis (Chapter 11): Department of Sociology, Memorial University, St. John's, Newfoundland A1C 5S7, Canada

Allenby T. Pinhorn (Chapter 4): Science Branch, Department of Fisheries and Oceans, Northwest Atlantic Fisheries Centre, Post Office Box 5667, St. John's, Newfoundland A1C 5X1, Canada

Brian J. Rothschild (Chapter 6): University of Massachusetts - Dartmouth 285 Old Westport Road, North Dartmouth, Massachusetts 02747-2300, USA

Fredric M. Serchuk (Chapter 2): National Marine Fisheries Service, Northeast Fisheries Science Center, Woods Hole, Massachusetts 02543, USA

Nancy L. Shackell (Chapter 8): Department of Oceanography, Dalhousie University, Halifax, Nova Scotia B3H 4J1, Canada

Alexei F. Sharov (Chapter 6): University of Massachusetts - Dartmouth, 285 Old Westport Road, North Dartmouth, Massachusetts 02747-2300, USA

Alan F. Sinclair (Chapter 3): Department of Fisheries and Oceans, Gulf Fisheries Centre, Post Office Box 5030, Moncton, New Brunswick E1C 9B6, Canada

Robert L. Stephenson (Chapter 13): Department of Fisheries and Oceans, St. Andrews Biological Station, St. Andrews, New Brunswick E0G 2X0, Canada

Ralph E. Townsend (Chapters 10 and 12): Professor of Economics, Department of Economics, University of Maine, 5774 Stevens Hall, Orono, Maine 04469-5774, USA

James A. Wilson (Coeditor): Department of Resource Economics, University of Maine, Orono, Maine 04469, USA

BOOK CONVENTIONS

This book has been written and edited by fishery professionals with two principal objectives, as the Preface has suggested. One was to treat the northwest Atlantic groundfish fishery and its dynamics as accurately and dispassionately as possible, an objective fostered by peer review. The other was to make the information—some of it highly technical—comprehensible to interested nonspecialists. Beyond the attention given to prose clarity, several conventions were adopted to keep the book from being a standard scientific treatise.

One convention was to separate the problem of fishery collapse from potential solutions of that problem. Part I of the book lays out the environmental and institutional contexts of the fishery, the history and current status of groundfish stocks, the cause(s) of modern stock declines, and the human consequences of those declines. Together, the chapters in Part I provide a synthesis of what is known and verifiable about the current fishery and its antecedents. With the factual groundwork thus laid, Part II turns to the more speculative future with a series of options for managing sustainable fisheries once stocks have recovered. Although rationally presented and not all mutually exclusive, these options are subject to debate among fishery stakeholders in the coming years.

Scientific and technical writing often is burdened by obligatory details important to professional readers. Many of these have been removed from chapter narratives in this book. Scientific names of fish and other species are listed for the only time on page xxi. Literature citations are restricted to the notes following each chapter, from where they are referenced (by the name-and-year system) to a comprehensive bibliography near the book's end. The notes themselves are identified in text by parenthetic numbers. The notes also are used to provide background or supplementary information about points made in the main texts.

Every effort has been made to define or explain in context the technical terms used in the book. For ease of reference, however, a glossary of key terms is provided at the back of the book.

Metric units of measure are used throughout the book. In particular, all tonnages given for landings or biomass are metric tons (1,000 kg; 2,200 pounds). The choice between metric and English systems was not easy. Metric is the official system of Canada and the standard system of science; English is the official system of the United States and the de facto basis of many fishery regulations even in Canada. English measures are given parenthetically when they will help clarity or precision. The relevant measurement equivalencies follow.

1 meter (m) = 3.28 feet = 1.09 yard = 0.55 fathom
1 kilometer (km) = 0.62 mile = 0.54 nautical mile
1 square kilometer = 0.39 square mile
1 degree Celsius (°C) = 1.80 degree Fahrenheit

Finally, the following national abbreviations are used without definition elsewhere in the book: UK for United Kingdom; U.S. for United States (adjective); USA for United States of America (noun); USSR for the former Union of Soviet Socialist Republics.

SPECIES LIST

Species of animals are cited by their common names throughout this book.[a] Their corresponding scientific names are given below. Species marked with an asterisk (*) make up the group known as "principal groundfish" for their importance in Canadian and U.S. fisheries.

Fish

Acadian redfish*[b]	*Sebastes fasciatus*
alewife	*Alosa pseudoharengus*
American plaice	*Hippoglossoides platessoides*
Arctic cod	*Boreogadus saida*
Atlantic cod*	*Gadus morhua*
Atlantic halibut	*Hippoglossus hippoglossus*
Atlantic herring	*Clupea harengus*
Atlantic mackerel	*Scomber scombrus*
Atlantic menhaden	*Brevoortia tyrannus*
butterfish	*Peprilus triacanthus*
capelin	*Mallotus villosus*
cusk	*Brosme brosme*
deepwater redfish*[b]	*Sebastes mentella*
fourspot flounder	*Paralichthys oblongus*
golden redfish*[b]	*Sebastes norvegicus*
goosefish	*Lophius americanus*
Greenland halibut[c]	*Reinhardtius hippoglossoides*
haddock*	*Melanogrammus aeglefinus*
longhorn sculpin	*Myoxocephalus octodecemspinosus*
lumpfish	*Cyclopterus lumpus*
ocean pout	*Macrozoarces americanus*
orange roughy[d]	*Hoplostethus atlanticus*
Pacific halibut	*Hippoglossus stenolepis*
Pacific salmon	Genus *Oncorhynchus*
pollock*	*Pollachius virens*
redfish	Genus *Sebastes*
red hake*	*Urophycis chuss*
scup	*Stenotomus chrysops*
searobins	Family Triglidae
silver hake*	*Merluccius bilinearis*
skates	Family Rajidae
sole	*Solea solea*
spiny dogfish	*Squalus acanthias*

striped bass	*Morone saxatilis*
summer flounder	*Paralichthys dentatus*
swordfish	*Xiphias gladius*
thorny skate	*Raja radiata*
weakfish	*Cynoscion regalis*
white hake	*Urophycis tenuis*
winter flounder	*Pleuronectes americanus*
winter skate	*Raja ocellata*
witch flounder	*Glyptocephalus cynoglossus*
yellow perch	*Perca flavescens*
yellowtail flounder	*Pleuronectes ferrugineus*

Crustaceans

American lobster	*Homarus americanus*
northern shrimp	*Pandalus borealis*
red king crab	*Paralithodes camtschaticus*
snow crab	*Chionoecetes opilio*

Mollusks

hard-shell clam[e]	*Mercenaria mercenaria*
sea scallop	*Placopecten magellanicus*
soft-shell clam[f]	*Mya arenaria*

Mammals

grey seal	*Halichoerus grypus*
harp seal	*Phoca groenlandica*

---

[a]Species names are those given by Robins et al. (1991a, 1991b) for fishes, Williams et al. (1989) for crustaceans, Turgeon et al. (1988) for mollusks, and Jefferson et al. (1993) for marine mammals. Complete citations are given in the References at the back of this book.

[b]The three species of redfish in the northwest Atlantic are somewhat difficult to tell apart, and they often are lumped as "redfish" in landings records and as redfish or "ocean perch" in trade. Both the Acadian and deepwater redfishes are also known as beaked redfish.

[c]Greenland halibut is also known as Greenland turbot.

[d]The name orange roughy often is applied to additional species of roughy that occur in commercial catches.

[e]An "official" name for this species is northern quahog.

[f]An "official" name for this species is softshell.

PART I

THE ONCE AND PRESENT FISHERY

Groundfish stocks of Atlantic Canada and New England supported one of the world's great fisheries for centuries. Now many stocks are decimated, and the fishery has been closed over much of the region. What happened?

A fishery comprises not only the fish exploited but the environment in which they live and the people who harvest them. Successful fishery management rests on an understanding of each component and of the interactions each component has with the others. Much is known about the northwest Atlantic groundfish fishery; much remains to be learned. Part I of this book summarizes knowledge of the fishery that was and the fishery that is today. Chapters move from an overview of the fishery's environmental context to the history and status of groundfish stocks and then to a close analysis of forces behind the fishery's decline. After a review of the institutional arrangements under which the fishery has been managed, Part I concludes by evoking the profound social consequences of a contracting fishery.

Understanding of the groundfish fishery was not strong enough to prevent the fishery's collapse. It is strong enough in retrospect to establish, with considerable assurance, the principal cause of the decline.

Ken F. Drinkwater and David G. Mountain

CLIMATE AND OCEANOGRAPHY

The abundance, growth, and geographical range of many fish stocks vary in response to environmental changes. The environment can affect fish both directly by altering their physiology and indirectly by altering the abundance of their food, predators, diseases, and parasites (1). Because most of these relations are not yet understood, fisheries managers generally have not considered environmental factors; they have focused instead upon the effects of fishing on spawning stock abundance. It is clear, however, that if fish stocks are to be sustained over the long term, we must better understand the environment, its changes, and the effects these changes have on fish survival and abundance. In this chapter we describe the main physical oceanographic features on the continental shelves along the northwest Atlantic and their year-to-year variability. We also discuss some of the possible effects this variability has on fish stocks. Our intent is to provide context and background for the considerations of groundfish stocks that follow in this book (2).

GEOGRAPHY

Catches of northwest Atlantic groundfish have centered historically on the extensive bank system off southeastern Canada and New England, but the fishery as a whole exploits the continental shelf from Hudson Strait in the north to Cape Hatteras in the south (Figure 1.1). The continental shelf in this region has an area of approximately 1,200,000 km^2. It varies in width from about 50 km off Cape Hatteras, North Carolina, to over 500 km on the Grand Bank, Newfoundland. Its principal named features include, from north to south, the Labrador Shelf, the northeast Newfoundland Shelf, the Grand Bank, the Gulf of St. Lawrence, the Scotian Shelf, the Gulf of Maine, and the Middle Atlantic Bight. The shelf topography is generally complex with numerous outer banks and inner basins. The banks

4   Climate and Oceanography

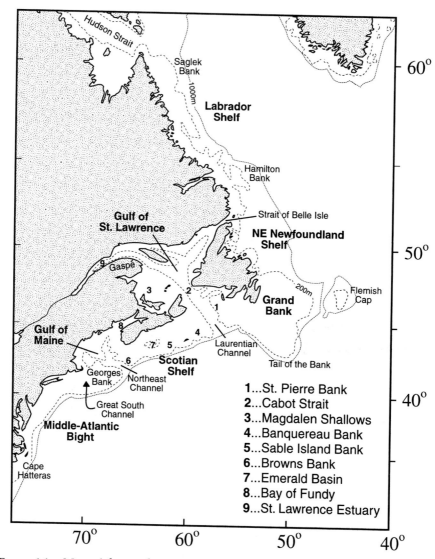

FIGURE 1.1.—Map of the northwest Atlantic groundfish areas. Depth contours are denoted by dashed lines; the 200-m-deep bottom contour outlines the principal banks and channels on the continental shelf and the 1,000-m contour lies over the continental slope.

are often separated by cross-shelf channels or saddles. The tops of the banks are typically 100–200 m deep, but they may be only 20–40 m deep in some locations. Depths of the channels and saddles usually exceed 250 m. The inner basins reach maximum depths of over 800 m on the Labrador Shelf, but they are typically 250–400 m deep. The major banks include Saglek and Hamilton off Labrador; Grand Bank, Flemish Cap, and St. Pierre Bank off Newfoundland; Banquereau, Sable, and Browns banks on the Scotian Shelf; and Georges Bank in the Gulf of Maine. All are, or have been, home to important commercial groundfish stocks.

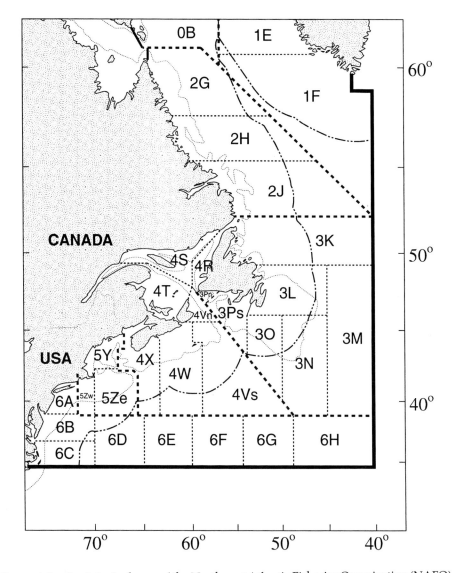

FIGURE 1.2.—Statistical scheme of the Northwest Atlantic Fisheries Organization (NAFO). The NAFO "convention area" is bounded by the heavy solid line; the area extends northwestward off the map. Subareas are numbered and separated by heavy broken lines. Divisions (capital letters) and subdivisions (lowercase letters) are separated by heavy dotted lines. The 200-m bottom contour is denoted by a light dotted line and the 370-km (200-nautical-mile) fishing boundary by a dashed–dotted line.

The deepest extensive shelf region (generally greater than 300 m) lies northeast of Newfoundland, whereas the shallowest (typically less than 100 m) is the Middle Atlantic Bight. The Middle Atlantic Bight is relatively smooth with no prominent banks, basins, or cross-shelf channels. Farther north, the major cross-shelf topographic features are the Laurentian Channel into the Gulf of St. Lawrence and the Northeast and the Great South channels into the Gulf of Maine. At the seaward edge of the shelf, the continental slope falls off rapidly to ocean basin depths of 2,000–4,000 m.

Fisheries catch statistics and stock identification are generally reported according to the Northwest Atlantic Fisheries Organization (NAFO) subareas. Our region of interest encompasses NAFO subareas 2 to 6 (Figure 1.2).

ATMOSPHERIC CONDITIONS

Air temperatures decrease northward and exhibit large seasonal changes. On average, winter coastal air temperatures typically vary from 10°C in the south to −25°C in the north and are below 0°C from the Gulf of Maine northwards. In summer, they range from 26°C in the south to less than 10°C in the north. Maximum winds occur during the winter, principally from the northwest, and generally weaker southwest or west winds blow in summer. Northwest winds bring cold, dry air from arctic and subarctic regions, whereas southwest winds carry warm air. Occasionally, easterly winds push cool, damp air in from the northern Atlantic Ocean.

In winter, the large temperature contrast between the cold continental and the warm oceanic air produces an atmospheric frontal zone off the east coast of North America between Florida and southern Newfoundland. Atmospheric conditions in this zone promote the formation of low-pressure storms, which then track northeastward along the front. The storm tracks tend to converge and stagnate in the vicinity of Iceland, resulting in the formation of a semipermanent "Icelandic Low" in surface barometric pressure. These storms carry relatively mild, moist air and can produce large temperature variations during the winter over the continental shelf regions north of Cape Hatteras. In summer, continental warming greatly decreases the atmospheric temperature gradient and weakens the atmospheric front off the coast, resulting in weaker, less-frequent storms in the northwest Atlantic. In late summer and early autumn, however, cyclones can develop over the tropical Atlantic or the Gulf of Mexico. These often intensify as they move northward along the Atlantic seaboard of the United States, and a few develop into extratropical storms (winds between 17 and 33 m/s) or hurricanes (winds exceeding 33 m/s).

The large-scale atmospheric circulation over the northern Atlantic is dominated by the Icelandic Low in tandem with a high pressure system that tends to be centered over the Azores, islands in the eastern Atlantic near the latitudes of Washington, D.C., and Lisbon, Portugal. When barometric pressure in the central Icelandic Low decreases, pressure in the Azores High increases, and vice versa. This dynamic linkage is known as the North Atlantic Oscillation (NAO). The NAO index is the difference in winter (December, January, February) sea level pressure between the Azores and Iceland (3). A high positive index (particularly high pressure in the Azores, particularly low pressure in Iceland) indicates strong westerly winds across the northern Atlantic and strong northwesterly winds into the Labrador Sea (4). As shown later, many atmospheric and oceanic changes on continental shelves in the northwest Atlantic are linked to variations in the NAO index.

Heat exchange with the atmosphere is one of the two major sources of regional oceanic temperature variability, the other being advection (transportation) of heat by ocean currents. The ocean is heated by energy from the sun, but it also loses heat by

Heat exchange with the atmosphere is one of the two major sources of regional oceanic temperature variability, the other being advection (transportation) of heat by ocean currents. The ocean is heated by energy from the sun, but it also loses heat by emitting long-wave back-radiation back to the atmosphere. Evaporation takes heat away from water and condensation of atmospheric moisture gives heat back; this exchange is called "latent heat flux." If the atmosphere is warmer than the ocean, heat is transferred to the ocean and vice versa; this exchange is called "sensible heat flux." Both latent and sensible heat fluxes increase with higher wind speed. The amount of air–sea heat exchange and the corresponding amount of warming or cooling of the ocean depend upon the local balance between solar heating, radiation of heat emitted by the ocean, and the latent and sensible heat fluxes. In the northwest Atlantic, shelf waters gain heat during spring and summer because of the dominance of solar heating, and they lose heat in winter due to increased evaporation and to the large difference between the relatively warm ocean and the cold air. Annually, shelf waters south of Newfoundland gain enough heat from the atmosphere to raise the temperature by 1–3°C (5). This increase does not actually occur because cold subarctic waters transported into the region compensate for the excess heating. Off the continental shelf south of the Grand Bank, the opposite occurs; there is a net annual loss of heat from the ocean to the atmosphere, heat that is replaced by advection of warm Gulf Stream waters.

RIVER RUNOFF

River runoff reduces the salinity and density of near-surface shelf waters and increases stratification, which is the difference in water density between the surface and subsurface layers. Large quantities of freshwater flow onto the continental shelves between Hudson Strait and Cape Hatteras, but the largest contribution comes from the St. Lawrence River system. The annual St. Lawrence discharge (424 km^3/year) exceeds the combined discharge of all rivers along the Atlantic United States south to Florida (6). Effects of the St. Lawrence River are felt throughout the western Gulf of St. Lawrence, along the Scotian Shelf, and into the Gulf of Maine. Other important rivers are the Churchill of southern Labrador, the Saint John draining into the lower Bay of Fundy, and the Hudson, Delaware, and Chesapeake Bay rivers emptying onto the Middle Atlantic Bight. In addition, large freshwater volumes discharging from Hudson Bay and James Bay tributaries flow through Hudson Strait onto the northern Labrador Shelf. The flow in most rivers exhibits a large seasonal range, peaking in spring as the winter snows melt and falling to a minimum in summer. Southern rivers tend to peak earlier than those further north.

SEA ICE AND ICEBERGS

Cold air and strong northwest winds in winter cause sea ice to form over the continental shelf from Hudson Strait to the Gulf of St. Lawrence (Figure 1.3). Ice normally forms on the northern Labrador Shelf by early December, spreads rap-

8   Climate and Oceanography

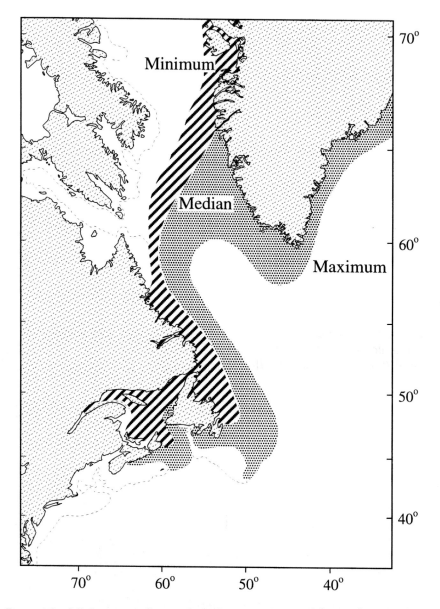

FIGURE 1.3.—Minimum, median, and maximum positions of the southern boundary of sea ice in the northwest Atlantic Ocean over the years 1953–1987. Sea ice reaches its southern-most extent in February, March, or April each year. Broken lines in this and subsequent maps denote the 200-m depth contour.

idly southward, and reaches the Strait of Belle Isle by late December. Ice also forms in the western Gulf of St. Lawrence by late December. By the end of February, ice has pushed southward to the Grand Bank and covered the Gulf of St. Lawrence. At its maximum seasonal extent, typically in March, ice may cover much of the Grand Bank and large portions of the Scotian Shelf. Ice then begins to retreat, and by mid-April the Gulf of St. Lawrence is generally ice free. By mid-May the ice has retreated to the southern Labrador Shelf and by mid-July it has disappeared from the Labrador Shelf altogether.

Sea ice modifies the air–sea heat exchange in several ways: it keeps solar energy from reaching the water, reflecting significant amounts back to the atmosphere; it reduces latent and sensible heat losses from the water; and it retards springtime warming of the ocean, taking up solar energy for its own melting. Meltwater from ice reduces the salinity of surface waters (7) and affects the position and strength of the shelf circulation, including the Labrador Current, by raising sea level on the shelf above levels farther offshore (8).

Ice also appears along the Labrador Shelf and onto the Grand Bank in the form of icebergs. Icebergs originate from glaciers on west Greenland and northern Baffin Island, drift around Baffin Bay for 1 or 2 years, and eventually move southward with the Labrador Current. In years when cold northwest winds are strong, icebergs are pushed further south and their melting is delayed. Over 90% of the icebergs passing onto the Grand Bank do so between March and July.

WATER CIRCULATION

Ocean currents tend to meander, eddy, diverge, and rejoin, so water circulation on and along the continental shelves of the northwest Atlantic is complex. The overall ("net" or "residual") flow carries cold, low-salinity waters towards the south (9). The Labrador Current coalesces on the northern Labrador Shelf from two sources (Figure 1.4). Waters from Hudson Strait form the inshore Labrador Current, over the shelf. At the continental slope, waters of west Greenland origin form an offshore branch, contributing approximately 80% to the transport of the Labrador Current (10). Upon reaching the southern Labrador Shelf, some of the flow is diverted into the Gulf of St. Lawrence through the Strait of Belle Isle (and other water is received in the reverse direction) while the remainder continues southward over the northeastern Newfoundland Shelf. The inshore branch rounds eastern Newfoundland, flows over the western Grand Bank and heads towards the Gulf of St. Lawrence. The offshore branch continues to the northeastern Grand Bank and then splits; some water flows eastward around the northern side of Flemish Cap but most of it turns southward along the eastern edge of the Grand Bank, around the Tail of the Bank, and westward to the Laurentian Channel. During its loop around the Grand Bank, the offshore branch loses substantial parts of its flow to adjacent offshore or "slope" waters.

The waters in the Gulf of St. Lawrence circulate counterclockwise, being joined on route by Labrador Current waters flowing in through the Strait of Belle Isle and by low-salinity waters emanating from the St. Lawrence estuary. Discharge from the estuary forms the Gaspé Current, which slows down as it spreads over the Magdalen Shallows and then strengthens as it narrows and exits the Gulf of St. Lawrence on the south side of Cabot Strait. As the water flows onto the Scotian Shelf, some rounds Cape Breton to form the Nova Scotia Current on the inner half of the shelf and the remainder continues along the Laurentian Channel before turning southwest at the shelf break. The latter branch is temporarily diverted inshore, flowing counterclockwise around Emerald Basin before heading back off-

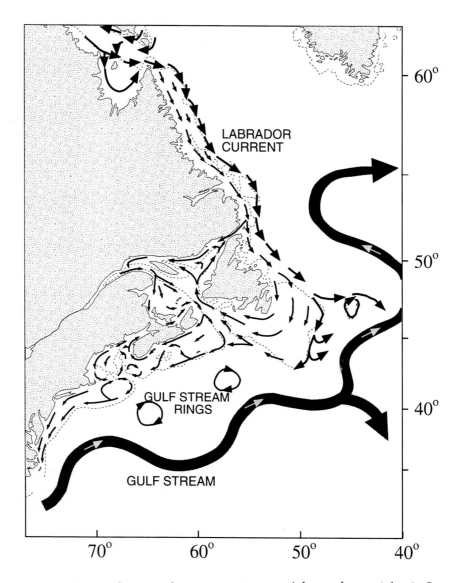

FIGURE 1.4.—Principal near-surface current systems of the northwest Atlantic Ocean. Heavier arrows mean larger volumes of transported water.

shore. Clockwise eddies form over several of the Scotian Shelf banks, including Browns Bank and Sable Island Bank. The Nova Scotia Current turns into the Gulf of Maine, part of it diverting into and around the Bay of Fundy and along the Maine coast. Upon reaching Cape Cod this current again splits. Part leaves the Gulf of Maine through the Great South Channel between Georges Bank and the Middle Atlantic Bight, but most moves northeastward, where its speed intensifies along the northern flank of Georges Bank. This flow also splits, one branch moving north to complete the Gulf of Maine gyre and the other flowing around Georges

Bank to form a clockwise gyre. In the Middle Atlantic Bight, the shelf flow continues its southwestward journey towards Cape Hatteras; maximum currents and greatest water transport occur near the shelf break.

Superimposed on the circulation just described are the tides, which result mainly from the approximately twice-daily rise and fall of sea level. Tidal currents typically flow at rates of 0.1 to 0.3 m/s, but they can exceed 1 m/s in the Gulf of Maine and the Bay of Fundy, where the range between high and low tides is particularly large. Where tidal currents flow over shallow areas such as Georges Bank or Nantucket Shoals south of Cape Cod, the water is intensely mixed top to bottom (11). Around banks with strongly sloping sides, such as Georges and Browns, tidal currents generate mean clockwise currents (12). Such flows act to retain the water on the banks by keeping it in place longer than otherwise (that is, the currents increase the "residence time" of bank waters). Circular currents have this effect wherever they occur.

From Cape Hatteras to the Tail of the Grand Bank, the offshore area between the northeast-flowing Gulf Stream and the southwest-flowing shelf currents is occupied by "slope waters." These waters tend to move relatively slowly in a counterclockwise direction. Large and variable meanders develop along the Gulf Stream and occasionally these meanders break off from the Gulf Stream to form rings or eddies. Rings forming on the north side of the Gulf Stream rotate clockwise (Figure 1.4) and trap warm Sargasso Sea water from south of the Stream into their centers, giving rise to the term "warm-core" rings. They have important influences on fishery dynamics, as we discuss later. Typically 20 to 30 warm-core rings form annually; although some persist for over a year, they usually only last a few weeks or months before dissipating or being reabsorbed by the Gulf Stream. They normally move slowly westward through the slope water and may "bump up" against the continental shelf. They are prevented from penetrating onto the shelf because their depth is much greater than the shelf (rings typically extend to depths exceeding 1,000 m) and the upper layers cannot separate from the lower layers because of dynamical constraints. The rings do, however, draw or "entrain" large amounts of water from the shelf into the slope water region; compensating flows of slope water move onto the shelf, usually at considerable depths through gullies and channels. Shelf waters that are advected offshore eventually lose their identity through mixing with slope waters.

The above description of the water circulation is based upon many years' observations (of temperature and salinity distributions and ocean currents) and on models of ocean movement. The currents at any given location, at any particular time, may differ markedly from those described because of influences of wind storms and seasonal or year-to-year changes. The long-term average southward flow of shelf water is, however, well established. It reveals the potential linkages between regions and the importance of upstream events in determining downstream conditions.

TEMPERATURE AND SALINITY DISTRIBUTIONS

Because of the large annual variability in atmospheric heat fluxes, shelf waters in the northwest Atlantic undergo large seasonal temperature changes. Indeed, the amplitude of the annual cycle in sea surface temperature there is the largest anywhere in the Atlantic Ocean (13) and one of the largest in the world. As described above, the ocean gains heat from the atmosphere in spring and summer through solar heating and loses heat in winter through latent (evaporative) and sensible heat fluxes. The heat gains and losses are then redistributed horizontally and vertically through advection and mixing. Salinities also undergo large seasonal variations, especially in areas influenced by freshwater inflows and the melting and freezing of ice. Mass exchanges between warmer, saltier offshore waters and cooler, fresher inshore waters lead to a net gain of heat and salt on the shelf.

On average, winter sea surface temperatures range from 6–10°C or more off Cape Hatteras to below 0°C under or near sea ice. Offshore, slope water temperatures range from 10–18°C at Cape Hatteras to 2–4°C in the Labrador Sea. Much larger variability occurs in summer: surface temperatures range from 26°C off Cape Hatteras to 2–4°C on the northern Labrador Shelf (Figure 1.5). The low surface temperatures in summer on the Labrador Shelf reflect the influence of the tidally mixed waters from Hudson Strait. The general pattern of increasing temperatures southward over the continental shelves is interrupted in the vicinity of the Gulf of Maine, where localized areas of lower temperatures are observed—for example, over Georges Bank and off southern Nova Scotia. These areas are subject to strong tidal currents that mix cool bottom waters into the surface waters. The annual range in sea surface temperature is approximately 4–6°C in northern Labrador (14) and 15–20°C in the southern Gulf of St. Lawrence, the Scotian Shelf, and the Middle Atlantic Bight.

Bottom temperatures generally show little or no seasonal variation below 150–200 m. North of the Scotian Shelf, bottom temperatures are typically less than 4°C except in the deep basins and troughs, where they may rise to 4–6°C (Figure 1.6). On the Scotian Shelf, temperatures change from 2–3°C in the northeast to over 8°C in the deep basins of central and southwestern areas. The deep basins contain offshore waters that penetrate onto the shelf through channels and gullies; cold water near bottom on the northeast Scotian Shelf reflects a lack of slope water penetration due to topographic barriers. High bottom temperatures indicative of slope water penetration are also observed in the Gulf of Maine. In the shallow Middle Atlantic Bight, high seasonal temperature variations arise principally from atmospheric heating.

Surface salinities generally increase seaward and show local minima where large rivers, such as the St. Lawrence, empty onto the continental shelves (Figure 1.7). In summer, surface salinities generally are less than 33 practical salinity units (psu) on the shelf, and they are lowest in the St. Lawrence estuary and on the Magdalen Shallows in the Gulf of St. Lawrence. Surface salinities in winter show a similar pattern to that in summer but typically are higher by 0.5–1 psu.

The vertical structure of the water column also undergoes a large seasonal variation. In winter, strong winds and cool air temperatures result in a large loss of heat from the ocean and the water is mixed down to depths of about 150 m or so over the shelf. In spring and summer, solar heating together with the runoff of freshwater makes the surface waters less dense and leads to rapid stratification of the water column, trapping the cold, winter-mixed water below. In many shelf areas, relatively warm water of offshore origin moves in along the bottom beneath the winter-cooled waters; although warmer, the offshore waters are more saline

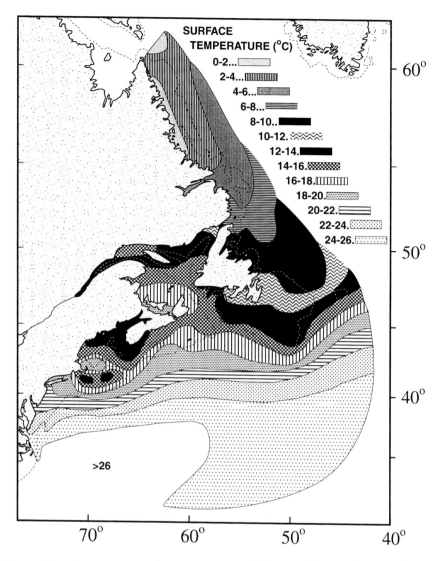

FIGURE 1.5.—Average sea surface temperatures (degrees Celsius) along northwest Atlantic shelves and slopes during August.

and hence more dense. Sandwiched between warmer layers, the winter-chilled waters are referred to as the cold intermediate layer (CIL). The absolute temperatures within the CIL vary from less than 0°C on the Labrador Shelf to 3–5°C on the Scotian Shelf and in the Gulf of Maine.

YEAR-TO-YEAR VARIABILITY

In the last several sections we have described the "average" atmospheric, sea ice, and oceanic conditions. In any year, conditions may vary substantially from the average. In this section we examine year-to-year variability.

14 Climate and Oceanography

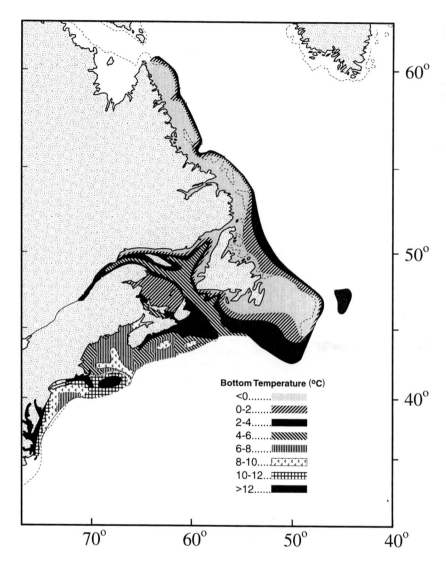

FIGURE 1.6.—Average near-bottom temperatures (degrees Celsius) along northwest Atlantic shelves and slopes during August.

Labrador and Newfoundland

Mean annual air temperatures at St. John's, Newfoundland, were relatively low from 1880 to the early 1900s, rose sharply in the late 1920s, peaked in the 1950s and remained high through the 1960s (Figure 1.8). Since then, temperatures have declined gradually and also fluctuated with a period of approximately 10 years; minima occurred in the early 1970s, the mid-1980s, and the early 1990s. Similar trends have been observed along the Labrador coast at Cartwright (data from 1938 to present) and on Baffin Island at Iqaluit (1947 to present). Cold temperatures occur in years when northwest winds push arctic air masses farther

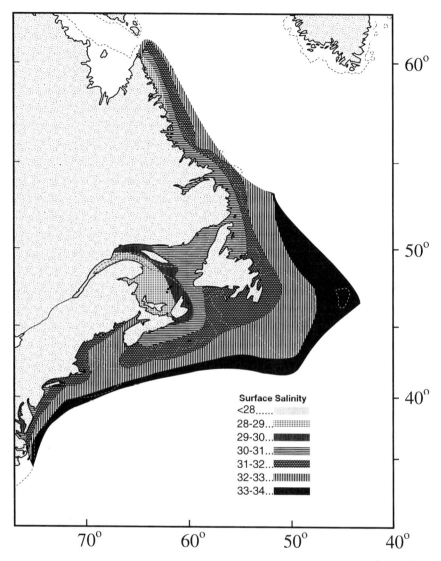

FIGURE 1.7.—Average sea surface salinities (practical salinity units) along the northwest Atlantic shelves and slopes during August.

south (Figure 1.9). The strong northwest winds result from a deepening of the Icelandic Low, which in turn produces an increase in the NAO index (Figure 1.9). Conversely, warm years occur when northwest winds and the Icelandic Low are weak and the NAO index is low.

Since the 1960s, when systematic records of offshore ice conditions began, sea ice off Labrador and Newfoundland has gradually been appearing earlier, lasting longer, and extending farther southward, consistent with decreasing air temperatures and increasing northwest winds (15). There were also distinct peaks in the area of sea ice that match the periods of coldest air temperatures and stron-

16   Climate and Oceanography

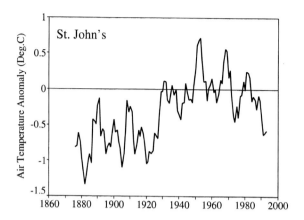

FIGURE 1.8.—Five-year running averages of annual air temperature anomalies (degrees Celsius) relative to the 1961–1990 mean (horizontal line) for St. John's, Newfoundland. An anomaly is a deviation from a specified value (here, the 30-year mean temperature). Running averages (each point is the average of adjacent years) reduce the influence of extreme values.

gest northwest winds (Figure 1.9). The cold temperatures promote ice formation in winter and delay melting in spring, and the strong winds increase the area of ice coverage by pushing the ice farther south. The number of icebergs reaching the Grand Bank has also increased since the 1960s; larger numbers appear in years of extensive ice cover (16).

Ocean temperatures and salinities have been monitored since the late 1940s at station 27, 10 km east of St. John's, Newfoundland. Year-to-year temperature changes at station 27 are representative of trends from southern Labrador to the Grand Bank (17). Temperatures were near normal into the early 1960s, peaked in the late 1960s, and have been declining since (Figure 1.9). Sea temperature minima tend to occur at the same time as minimum air temperature and maximum northwest winds and sea ice extents. Sea temperatures during the first half of the 1990s were the coldest on record and followed a decade of below-normal temperatures. Also, the amount of cold water, measured by the areal extent of the CIL (waters colder than 0°C), has increased (Figure 1.9). Salinities show no overall trend since the late 1940s but were high during the warm 1960s. There have also been three notable salinity minima since then, one in each decade. Those in the 1980s and 1990s corresponded in time to sea temperature minima. The 1970s salinity minimum preceded the temperature minimum by 1 to 2 years (18).

In years of cold air and sea temperatures, strong northwest winds, and high NAO index, the volume of water transported by the Labrador Current is low; in warm years with weak winds and a low NAO index, the transport is high (19). Similar variability in the current off west Greenland (20) suggests a large-scale response of the Labrador Sea to atmospheric forcing, although the mechanism is not understood.

Local atmospheric heating cannot account for the observed year-to-year changes in sea temperatures at either station 27 or over the Grand Bank, but air–sea exchanges "upstream" likely play an important role (21). Temperature and

Climate and Oceanography 17

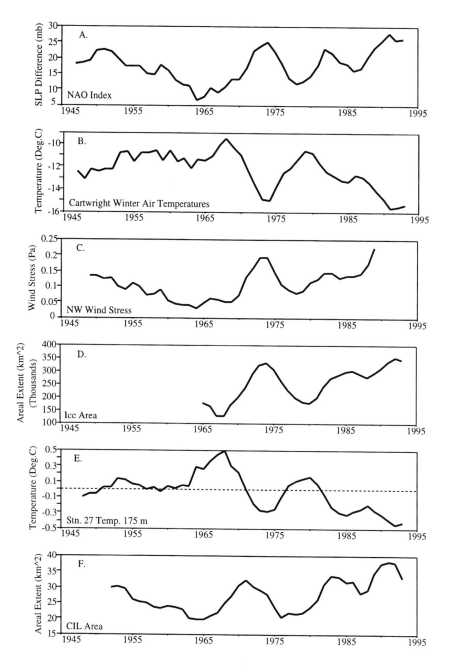

FIGURE 1.9.—Five-year running means of (**A**) the North Atlantic Oscillation (NAO) index, (**B**) winter air temperatures at Cartright, Labrador, (**C**) northwest wind stresses over the Labrador Sea, (**D**) the ice extent south of 55°N during February, (**E**) the near-bottom (175-m-deep) temperature anomalies at station 27 (Stn.) off Newfoundland, and (**F**) the areal extent of the cold intermediate layer (CIL) off Bonavista, Newfoundland, during summer, 1945–1995. Other abbreviations: SLP is sea level pressure; mb is millibars; Deg.C is degrees Celsius; Pa is pascals (a measure of pressure); NW is wind from the northwest; km^2 is square kilometers.

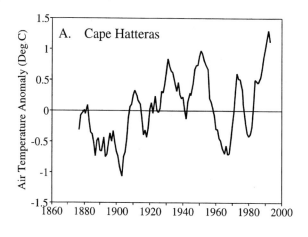

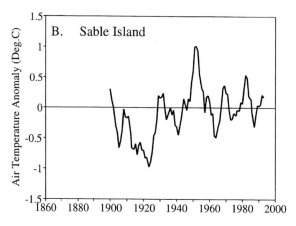

Figure 1.10.—Five-year running means of annual air temperature anomalies (degrees Celsius) relative to the 1961–1990 means at (**A**) Cape Hatteras, North Carolina, and (**B**) Sable Island, Nova Scotia.

salinity changes off Newfoundland have also been linked with hydrographic changes off west Greenland; events off Greenland appear to arrive off Labrador and Newfoundland 1 to 2 years later (22). Further studies are required to determine more precisely the role of atmospheric forcing and advection in controlling variability in the water properties on the Labrador and Newfoundland shelves.

Gulf of St. Lawrence to the Middle Atlantic Bight

Air temperatures from the Gulf of St. Lawrence to Cape Hatteras were below normal during the late 1800s and early 1900s, increased through the 1920s, and generally were above normal through to the 1950s, trends similar to those in Newfoundland and Labrador (Figure 1.10). Since then, and in contrast to the more northern region, air temperatures declined to near-historical lows by the mid to late 1960s, rose rapidly into the 1970s, and then oscillated near the long-term mean (off eastern Canada) or rose substantially more (southern Middle Atlantic Bight). At Cape Hatteras, warm years tend to occur when the NAO index is high (23). A higher NAO index means a strengthened Azores High, stronger southerly winds carrying warm tropical air masses to the Middle Atlantic Bight, and thus higher temperatures. This is opposite to the response in Newfoundland where a higher NAO index means a deeper Icelandic Low, stronger northwest winds, and colder temperatures over the Labrador Sea. In the Gulf of Maine and on the Scotian Shelf, which lie between the influence of the Low and the High, air temperatures appear to be unrelated to variations in the NAO index.

Water temperature and salinity have been monitored at Prince 5, a hydrographic station at the mouth of the Bay of Fundy, since 1924. The water there cooled and freshened from the early 1950s to the mid-1960s and then rapidly reversed these trends into the 1970s (Figure 1.11). This temperature pattern also has been observed in shelf waters from the Laurentian Channel to the Middle Atlantic Bight (24). The pattern cannot be explained by atmospheric heating (25); it originates instead from changes in characteristics of the slope water that is subsequently advected onto the shelf (26). Changes in slope water characteristics, in turn, are related to variations in the strength of the Labrador Current's deeper (100–300 m) westward flow along the continental slope.

During the past decade, waters in the deep basins and channels of the Scotian Shelf and Gulf of Maine have been relatively warm except for a brief appearance of cold water in the early 1990s. At the same time, significant cooling has been observed in the intermediate waters and shallow bottom waters from the northern Gulf of Maine to southern Newfoundland and in the Gulf of St. Lawrence. The emerging differences in temperature with depth contrasts with the nearly homogeneous temperatures of the 1940s to the mid-1980s (27). Temperatures in the CIL declined by approximately 1–2°C from the mid-1980s to the 1990s, falling in several areas to levels at or below those recorded during the cool 1960s (28). The cold intermediate layer extends to the bottom over most of the Magdalen Shallows and on the northeastern Scotian Shelf. The cold water is believed to be due to a combination of local winter cooling and advection from off northern Newfoundland and the Grand Bank.

In the Middle Atlantic Bight during the 1960s to the late 1980s, temperature variations were similar throughout the water column (29). The late 1960s was the coldest period and the mid-1970s the warmest; temperatures were intermediate during the late 1970s and early 1980s. The range from the coldest to the warmest temperatures was approximately 4°C. The salinity of the shelf waters (defined by salinities less than 34 practical salinity units) also exhibited large year-to-year variations (range, 1 psu) with minima in the late 1970s, 1984, and 1988 separated by years of relatively high salinity in 1981 and 1986. These changes reflected variations in local river discharge and precipitation and, to a minor extent, in the amount of freshwater that had left the St. Lawrence River 17 months previously (30). Changes in the inflow of slope water to the Gulf of Maine also affect the salinities in the Middle Atlantic Bight. Water in the Bight originates from the Gulf of Maine, where cold, low-salinity water flowing south over the Scotian Shelf mixes with relatively warm, saline slope water entering via the North-

20   Climate and Oceanography

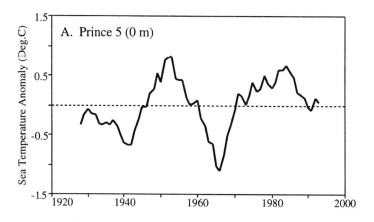

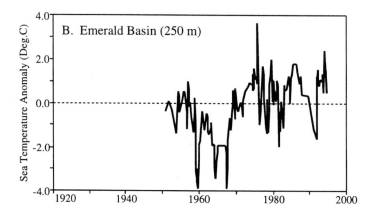

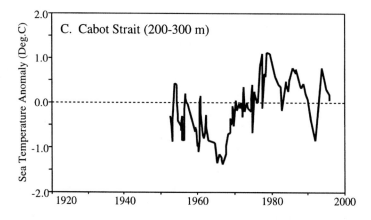

FIGURE 1.11.—Temperature anomalies (degrees Celsius) relative to 1961–1990 means (**A**) at the surface (0-m depth) at Prince 5 in the Bay of Fundy, (**B**) at 250-m depth in the Emerald Basin, and (**C**) in the 200–300-m layer in Cabot Strait. Prince 5 values are 5-year running means of annual average temperatures, whereas Emerald Basin and Cabot Strait values are plots of all available monthly means. The latter two curves have not been smoothed because data for many months are missing.

east Channel. The resultant water mass is eventually carried into the Middle Atlantic Bight, where it continues its journey southward. Increased inflows through the Northeast Channel are believed to have contributed to the warm and saline conditions in the Middle Atlantic Bight in the mid-1980s (31).

ENVIRONMENTAL CHANGES AND FISH

Fish respond to environmental change in many well-documented ways. Here, we provide a few examples of how variations in the oceanic conditions described in this chapter have affected northwest Atlantic groundfish, their prey, and their predators. Because most fish species or stocks prefer a specific temperature range (32), long-term changes in temperature can lead to shifts in the geographic ranges of the species. Such shifts are generally most evident near a species' northern or southern boundary; warming of the waters results in a distributional shift northward, and cooling draws species southward. For example, capelin, a cold-water species and the major food of Atlantic cod off Newfoundland and Labrador, spread southward as far as the Bay of Fundy when temperatures declined in the mid-1960s, but disappeared from the south as temperatures rose in the 1970s (33). During the cooling of the late 1980s and early 1990s, capelin again extended their range, moving eastward to Flemish Cap and southward onto the northeastern Scotian Shelf, where they now spawn successfully (34). Arctic cod also spread southward from Labrador, and groundfish occupying the southern Labrador and northern Newfoundland shelves such as Atlantic cod, Greenland halibut, American plaice, and several noncommercial species shifted their distributions further southward and offshore (35). Latitudinal shifts in response to temperature changes have also been documented for groundfish in the Gulf of Maine (36).

Catch rates can also be influenced by environmental conditions. For example, the catch of age-4 Atlantic cod during spring research surveys on the eastern Scotian Shelf is greater during years when a larger proportion of ocean bottom is covered by CIL waters with temperatures less than 5°C and salinities of 32–33.5 psu (37). The larger catches may reflect aggregations of cod seeking preferred conditions or a reduced ability of the fish to avoid trawls at lower temperatures (38).

Mean bottom temperatures are responsible for 90% of the observed (10-fold) difference in growth rates between different Atlantic cod stocks in the northern Atlantic (39). Warmer temperatures lead to faster growth rates. In the northwest Atlantic, the largest cod are found on Georges Bank, where a 4-year-old fish is five times bigger, on average, than one off Labrador and Newfoundland. Temperature not only accounts for the differences in growth rate between stocks but also for year-to-year changes in growth rates within a stock. Thus, temperature declines during the last decade are in part responsible for the observed decrease in size at age of Atlantic cod on the Scotian and Newfoundland shelves (40). This is particularly important in that 50–75% of the declines in spawning stock biomass of the Newfoundland, Gulf of St. Lawrence, and northeastern Scotian Shelf cod stocks during this period were caused by reduced weight at age (41). Temperature-dependent growth rates are not restricted to cod. Cold temperatures have reduced the growth of American plaice on the Grand Bank (42) and of capelin off Newfoundland (43).

Understanding the variability in recruitment of fish to exploitable stocks has been the number one issue in fisheries science this century. Changes in fish abundance in the absence of fishing suggest environmental causes. Since the advent of intensive fishing, it has become increasingly difficult to sort out the relative influences of fishing and environment on recruitment. Recruitment variability has frequently been associated with variations in temperature during the first years of life of the fish (44). Recruitment of Atlantic cod off west Greenland, Labrador, and Newfoundland generally increases when ocean temperatures are warm and decreases when temperatures are cold (45). However, the warm periods also have been the periods in which spawning stock biomasses were high, and thus temperature cannot be confirmed as the main cause of recruitment variability. During the last 10 years of extremely cold temperatures in the northern regions, recruitment of Atlantic cod has been poor from Labrador to the Grand Bank. At the same time, as previously mentioned, cod moved further southward. A recent hypothesis is that these two features are related; in cold years, spawning tends to occur at southerly locations where ocean circulation patterns do not favor retention of cod larvae in suitable habitats; hence survival is poor (46). Temperature also can affect recruitment through its influence upon spawning times and maturation rates. Cold temperatures prior to spawning slow gonad development and spawning is typically delayed; conversely, warm temperatures result in earlier spawning through faster gonad development. This relationship has been observed for Atlantic cod on the northern Grand Bank (47). Significant delays in spawning may mean that the young encounter poor food supplies, resulting in slow growth and high mortality, as demonstrated for capelin off Newfoundland (48). Adults also can be adversely affected by cold. In the laboratory, American plaice reduced their feeding and lost weight at low temperatures (49), and it has been suggested that a weakened condition or starvation resulting from poor feeding caused the observed reproduction failure of the late 1980s and early 1990s. In the Gulf of St. Lawrence, survival of Atlantic cod to recruitment is generally better when freshwater runoff from the St. Lawrence River system is above normal (50), possibly because nutrient levels and hence food production are greater in such years.

Eggs and larvae are affected by currents. The 1987 haddock year-class on eastern Georges Bank, which appeared to have been spawned normally in early spring, was almost entirely displaced to the Middle Atlantic Bight by June, the result of unusually great water transport to the south that year (51). Recent improvements in numerical models of the currents over the continental shelves have allowed scientists to study the potential drift patterns of eggs and larvae (52). Advection into unfavorable sites may lead to reduced recruitment. For example, recruitment to many groundfish stocks is reduced when several Gulf Stream rings move close to the continental shelf during the spawning or larval periods (53). The leading hypothesis is that rings entrain eggs and larvae and transport them into offshore waters, where they die because they cannot find appropriate habitat into which to settle or because they encounter temperatures that are too high (54).

Later chapters in this book document in detail the drastic decline in the abundance of many commercial groundfish species in the northwest Atlantic in recent years. In most areas, these declines occurred during a period when the ocean temperatures cooled significantly. Although the main cause of the declines is felt to be overfishing (55), the environment has played a role by its negative effects on the growth, distribution, and perhaps recruitment of groundfish species. Interactions between fishing and environmentally induced changes may also have occurred. For example, when low temperatures caused lower growth rates, more small fish may have been (illegally) dumped from vessels at sea. We believe that improved environmental conditions will be a necessary but not a sufficient condition of groundfish stock recovery. Better management of our groundfish resources in the future will require increased knowledge of groundfish responses to environmental change as well as to fishing pressure.

NOTES

1. Shepherd et al. (1984).
2. Groundfish are species that live most of their lives on or near the ocean bottom, but their eggs and larvae occur higher in the water column where they are affected by prevailing currents and other oceanographic conditions. More extensive and detailed reports on the region's oceanography have been published by Hachey et al. (1954), Petrie and Anderson (1983), Chapman and Beardsley (1989), Koutitonsky and Bugden (1991), Smith and Schwing (1991), and Loder et al. (in press).
3. Rogers (1984).
4. Wind tends to blow at approximately right angles to the atmospheric pressure gradient. In the northern hemisphere, wind blows away from an observer when high pressure is to the right (this is known as the geostrophic relationship). Because the NAO index is a measure of the north–south atmospheric pressure gradient, it determines the strength of the westerly winds over the North Atlantic Ocean. The northwest winds during the winter over the Labrador Sea are part of the counterclockwise atmospheric circulation pattern associated with the Icelandic Low pressure system. When the Icelandic Low intensifies, the NAO index increases and the northwest winds over the Labrador Sea become stronger.
5. Petrie et al. (1994).
6. Sutcliffe et al. (1976).
7. Myers et al. (1990).
8. Lazier and Wright (1993).
9. Hachey et al. (1954); Chapman and Beardsley (1989).
10. Lazier and Wright (1993).
11. Garrett et al. (1978).
12. Loder (1980); Greenberg (1983).
13. Weare (1977).
14. In ice-infested northern waters, the lower limit on surface temperature is set by the freezing point of salt water (near $-2°C$); the salt is responsible for lowering the freezing point below that of freshwater ($0°C$).
15. Prinsenberg and Peterson (1994).
16. Marko et al. (1994).
17. Petrie et al. (1992); Colbourne et al. (1994).
18. The salinity minima of the 1970s was associated with the "great salinity anomaly" (GSA). The GSA began in the Greenland Sea, when large amounts of ice were transported by unusual winds out of the Arctic. When this ice subsequently melted, its freshwater was advected around the Labrador Sea to the Grand Bank (Dickson et al. 1988).
19. Myers et al. (1989); Petrie and Drinkwater (1993).
20. Myers et al. (1989).

21. Umoh et al. (1995).
22. Dickson et al. (1988); Myers et al. (1988); Stein (1993).
23. Drinkwater (1996).
24. Lauzier (1965); Petrie and Drinkwater (1993).
25. Umoh (1992).
26. Petrie and Drinkwater (1993). That slope water could induce variability in the properties of deep waters on the shelf had been suggested earlier for the Laurentian Channel (Lauzier and Trites 1958; Bugden 1991), the Scotian Shelf (Lauzier 1965), and the Gulf of Maine (Colton 1968).
27. Petrie and Drinkwater (1993).
28. Gilbert and Pettigrew (1997); Drinkwater et al. (1996).
29. Mountain and Murawski (1992).
30. Manning (1991).
31. Mountain (1991).
32. Scott (1982).
33. Frank et al. (1996).
34. Nakashima (1996); Frank et al. (1996).
35. The southward movement of Arctic cod was suggested by Gomes et al. (1995) and has been substantiated by annual autumn groundfish surveys off Newfoundland throughout the 1990s (Lilly et al. 1994). Distributional changes have been documented for Atlantic cod (deYoung and Rose 1993; Taggart et al. 1994; Rose et al. 1994), capelin (Lilly 1994), and assemblages of both commercial and noncommercial species (Gomes et al. 1995).
36. During the cooling of the 1960s, American plaice moved southward (Colton 1972). Mountain and Murawski (1992) found significant positive relationships between bottom temperature and the mean latitude of the catch for 14 of 30 species caught in spring trawl surveys during 1968–1989 (i.e., catches increased when temperatures did). Species showing positive relationships included commercial groundfish such as yellowtail flounder and noncommercial species such as goosefish and ocean pout.
37. Smith et al. (1991).
38. Smith and Page (1996).
39. Brander (1995).
40. Approximately 50% of the decline in growth rates of Atlantic cod on the northeastern Scotian Shelf during the 1980s and early 1990s can be explained by decreasing temperatures (Campana et al. 1995). Growth rates of northern cod off southern Labrador and northern Newfoundland become lower when the area of cold intermediate layer waters, defined in this case by temperatures less than 0°C, is high (de Cárdenas 1996; Shelton et al. 1996).
41. Sinclair (1996).
42. Brodie (1987).
43. Nakashima (1996).
44. Drinkwater and Myers (1987).
45. West Greenland: Buch et al. (1994); Labrador and Newfoundland: Taggart et al. (1994).
46. deYoung and Rose (1993).
47. Hutchings and Myers (1994a). Warm temperatures promote gonad development, resulting in earlier spawning, but the relationship between temperature at the spawning site and time of spawning depends on local hydrography and fish distribution. For Atlantic cod, earlier warm temperatures lead to earlier spawning on the northern Grand Bank, but cold temperatures lead to earlier spawning off southern Newfoundland (Hutchings and Myers 1994a). Fish in the latter area reside in warm slope waters and

move on to St. Pierre Bank prior to spawning. In very cold years on the Grand Bank, southern fish remain longer in the warm slope waters, resulting in faster gonad development and an earlier readiness to spawn.

48. Nakashima (1996).
49. Morgan (1992).
50. Chouinard and Fréchet (1994).
51. Polacheck et al. (1992).
52. Werner et al. (1993); Lough et al. (1994).
53. Myers and Drinkwater (1989).
54. Yellowtail flounder larvae in shelf water died suddenly when they were drawn into warm slope water off Georges Bank (Colton 1959).
55. Hutchings and Myers (1994b); Serchuk et al. (1994a).

Steven A. Murawski, Jean-Jacques Maguire, Ralph K. Mayo, and Fredric M. Serchuk

GROUNDFISH STOCKS AND THE FISHING INDUSTRY

Important groundfish fisheries have occurred in the region from Newfoundland to New England for over 400 years of recorded history (1). Landings of Atlantic cod and other species were being reliably documented by the early 1800s. By the turn of the Twentieth Century, the groundfish fishery supported major domestic industries for salted and fresh fish as well as export trades (primarily for salt cod) with Europe, the Caribbean, and South America.

The first half of the Twentieth Century saw the development and increasing use of the otter trawl to catch cod and related groundfish, particularly in the United States. Otter trawling also proved very efficient at capturing flatfishes, too, and fisheries for these species expanded rapidly. (Although otter trawling was introduced into Canada early in the century, it was banned or severely limited in many areas until the late 1940s.) As methods of harvesting, handling, transportation, and marketing of groundfish improved into the 1930s, markets switched from primarily salted fish to fresh and frozen fish, and fisheries for species such as haddock and redfish developed to supply distant markets for these products. Groundfish fisheries developed rapidly in the early part of the century, and although a few cases of obvious overfishing arose (such as for haddock on Georges Bank), increased fishing for an ever-widening array of species and stocks generated adequate fish supplies.

After World War II, a build-up of "distant-water" fleets from Europe and Asia in the northwestern Atlantic resulted in greatly increased landings of groundfish, especially of Atlantic cod, silver hake, haddock, and redfish. Fishing intensified after 1960 in the (then) international waters beyond 22 km from shore. Total landings of principal groundfish expanded from about 1 million tons in the early 1950s to over 2 million tons by 1965. Soon thereafter, however, fish stocks declined sharply in spite of restrictive management regulations enacted through the

International Commission for the Northwest Atlantic Fisheries (ICNAF). Canada and the United States extended their exclusive fishery zones to 370 km (200 nautical miles) in 1977. (Both countries passed authorizing legislation in 1976.) In the years following extensions of jurisdiction, foreign fishing was curtailed, fish stocks began to recover, and both Canadian and U.S. fishing fleets were modernized and expanded. Fishing pressure on important stocks again increased rapidly, this time from domestic fishing, despite fundamentally different management philosophies and techniques in the two countries. By the early 1990s, landings of a majority of groundfish stocks throughout the region were once more declining rapidly. Severe government restrictions on groundfish harvests in both countries soon followed.

This chapter documents the development and status of fishing industries and stocks supporting groundfish fishing in the northwest Atlantic. We review the exploitation history of major stocks, particularly in the last half of the Twentieth Century, as well as current stock status, short-term prognoses, current management approaches, and measures enacted to protect and rebuild the depleted groundfish resource (2).

STATUS OF THE RESOURCES

The so-called "principal" groundfish species of the northwest Atlantic, as defined by the Northwest Atlantic Fisheries Organization (NAFO, an international management body that succeeded ICNAF in 1976), are Atlantic cod, haddock, pollock, redfish (three species), and red and silver hakes (see listing at the front of this book). Most of these species range from Labrador to Cape Cod, and they have contributed most of the historical groundfish landings in the region. Flatfishes (flounders and halibuts), a second category of groundfish, support average landings about one-third those of the principal groundfish, but they are extremely valuable to the region's fisheries. Other groundfish include a variety of species historically taken as bycatches in fisheries for principal groundfish and flounders. Many of these species (e.g., spiny dogfish, skates, goosefish, lumpfish, and white hake) have become the objects of targeted fisheries in recent years as traditional fisheries have declined. (Targeted species are those explicitly sought by fishers. Bycatch is the incidental catch of nontargeted fish in the gear used.)

Groundfish stocks in the northwestern Atlantic are defined to a large degree by the NAFO statistical division(s) in which they occur (Figure 2.1). As described later in this chapter, an international agreement (convention) established a northwest Atlantic management area that extends from northwestern Greenland (Denmark) and Ellesmere Island (Canada) south to Cape Hatteras (USA) and east to 42° west (the longitude of central Greenland). Within the convention area, numbers (increasing from north to south) designate statistical "subareas." To the far north, subareas 0 (the Canadian side of Baffin Bay and Davis Strait) and 1 (the Greenland side of these waters) sustain some fishing for groundfish, especially Greenland halibut very recently. Subareas 2 (Labrador) south to 6 (off the middle Atlantic states) have supported most historical groundfish harvests, and this is the region we address in this chapter (see Figure 1.1 for place names).

Groundfish Stocks and the Fishing Industry   29

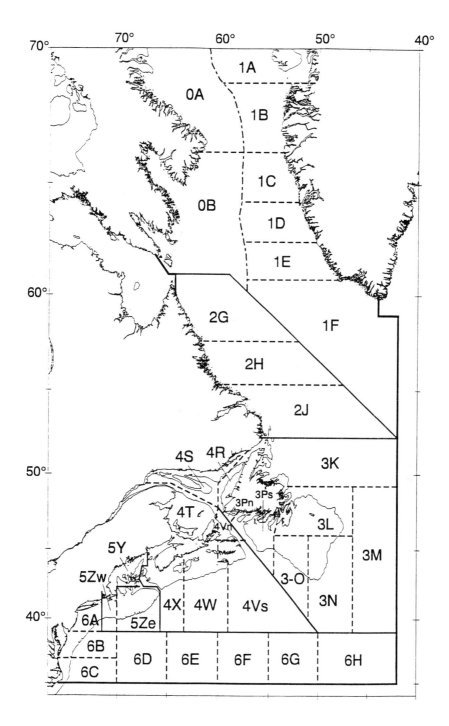

FIGURE 2.1.—Northwest Atlantic Fisheries Organization convention area (heavy borders), subareas (numbers), divisions (capital letters), and subdivisions (lowercase letters).

Within subareas, "divisions" are identified by capital letters (alphabetically from north to south, then starting over). Groundfish stocks are labeled by their divisions; thus Atlantic cod stock 3M refers to fish in subarea 3 (east and southeast of Newfoundland) and division M (around and south of the Flemish Cap; Figure 2.1). Stocks that cross statistical boundaries contain all relevant divisions in their names; stock 5YZ occurs in subarea 5 (south and east of New England) and in both divisions Y (Gulf of Maine) and Z (Georges Bank). If a stock crosses subareas, the crossing is noted by a + sign. Stock 2J+3KL occurs in subareas 2 (Labrador) and 3 (Newfoundland) and the relevant divisions thereof.

Divisions are further segregated into subdivisions, denoted by lowercase letters. Most subdivisions represent compass directions (e = east, s = south, etc.). In boundary areas, however, lowercase letters can designate national jurisdictions (c = Canada; u = USA).

The principal groundfish and flounder resources of NAFO subareas 2–6 (Labrador to Cape Hatteras) consist of 55 regulated stocks and 2 stock complexes (see the Appendix; this summary does not include red hake, which are not managed explicitly in the region). Atlantic cod has the largest number of regulated stocks—12—followed by redfishes with 8, haddock with 6, and American plaice with 6 stocks. Groundfish stocks within extended Canadian and U.S. jurisdictions are managed by domestic agencies. Stocks outside national jurisdictions are managed primarily by NAFO. Some stocks overlap jurisdictions (transboundary stocks) and may be managed differently in each (3).

Before the mid-Twentieth Century, total landings of principal groundfish averaged about 1 million tons. Most of the landings were taken by fishers from Newfoundland (a British dominion until 1949, when it became the 10th Canadian province), Canada, the United States, Spain, Portugal, and France. Landing records for all countries fishing in the region were not complete until after 1951. After the early 1950s, landings of principal groundfish rose quickly to a peak of 2.1 million tons in 1965. Much of this increase was fueled by greater catches first of cod and redfish and then of silver hake and haddock (Figure 2.2). Total landings began dropping after 1965 as stocks of silver hake, haddock, and red hake declined, and landings plummeted in the mid-1970s. Cod landings peaked in 1968 at 1.48 million tons, more than half of them being northern cod alone (4), and declined sharply thereafter. Between the peak of principal groundfish landings in 1965 and extensions of territorial jurisdictions in 1977, landings declined 64% to 760,000 tons (Figure 2.2). Principal groundfish landings then rose steadily, peaking again in 1986 at 1.05 million tons. After 1986, landings of principal groundfish plummeted to 337,000 tons in 1993, a 68% decline to the lowest level in this century.

Over the period 1952–1990, Atlantic cod accounted for 50–75% of annual principal groundfish landings; haddock, silver hake, and redfish made up the bulk of the rest (Figure 2.2). Since 1990, cod landings have fallen to about a third of the total, redfish and silver hake accounting for most of the remainder. In 1993, redfish made up about 40% of the landings of principal groundfish in the region, the highest percentage since 1952.

Landings of flatfishes, like those of principal groundfish, increased rapidly after the mid-1950s (Figure 2.3), a reflection of the increase in fishing activity by countries other than Canada and United States. Flatfish landings increased five-

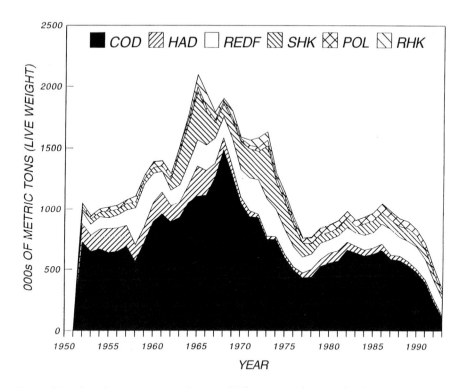

FIGURE 2.2.—Landings of principal groundfish species (thousands of tons) in NAFO subareas 2–6, 1952–1993. Species are Atlantic cod (COD), haddock (HAD), redfish (REDF, two or more species), silver hake (SHK), pollock (POL), and red hake (RHK).

fold from about 60,000 tons in 1955 to 315,000 tons in 1968. Much of this increase was attributable to American plaice, which accounted for 30–40% of annual flounder landings until the 1990s, but yellowtail and witch flounders and Greenland halibut (turbot) also contributed to the rapid rise. Following the peak in 1968, total flatfish landings decreased steadily to 109,000 tons in 1993, a 65% reduction from the maximum. Greenland halibut now account for about half of the total flatfish landings in the region; American plaice and yellowtail and witch flounders account for much of the rest (Figure 2.3). Most of the Greenland halibut landings come from a recently developed fishery in subarea 3.

After it extended national jurisdiction in 1977, the United States realized total landings of principal groundfish and flatfish that rose steadily to 218,000 tons in 1980; landings then declined 50% to 108,000 tons in 1993 (Figure 2.4). Canadian landings increased from 469,000 tons in 1976 to 812,000 tons in 1982, then declined 68% to 258,000 tons in 1993 (Figure 2.4). In the United States, the major decline occurred in 1985 and landings decreased slowly thereafter. In Canada, the decrease was gradual in the 1980s and precipitous in the 1990s.

In the following sections, we briefly recount the historical fishery trends and current status for all the regulated principal groundfish and flounder resources off eastern North America (NAFO subareas 2–6). The Appendix to this chapter summarizes the information in tabular form.

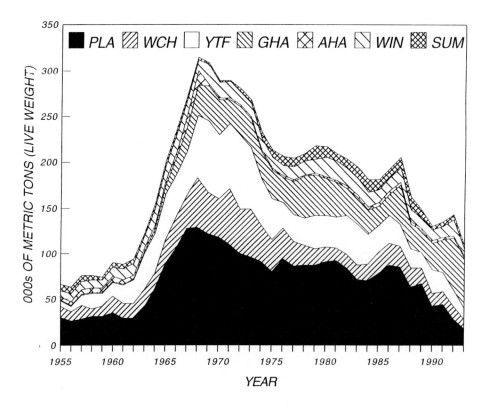

FIGURE 2.3.—Landings of flounders (thousands of tons) in NAFO subareas 2–6, 1952–1993. Species are American plaice (PLA), witch flounder (WCH), yellowtail flounder (YTF), Greenland halibut (GHA), Atlantic halibut (AHA), winter flounder (WIN), and summer flounder (SUM).

STOCK CHARACTERISTICS AND STATUS

Atlantic Cod

The Atlantic cod resource of eastern North America consists of 12 managed stocks or identified stock complexes (Appendix). Two stocks are managed primarily by NAFO (3M and 3NO, which are wholly or mainly beyond the 370-km zone of national jurisdiction), eight are under domestic control by Canada (2GH, 2J+3KL, 3Pn+4RS, 3Ps, 4TVn, 4Vn, 4VsW, 4X), one is under domestic control by the United States (5Y), and one is considered transboundary (5Z+6, portions of which are managed independently by Canada and the United States).

Cod landings in the region varied between 400,000 and 625,000 tons between the 1890s and 1950s, averaging about 500,000 tons (these figures do not include landings by Spain prior to 1952 [5]). During the first half of the Twentieth Century, cod were landed primarily by Newfoundland (about 50%); Canada and other countries also took substantial amounts, but U.S. landings declined as the fishery increasingly targeted haddock (Figure 2.5). Total landings increased rapidly after 1955, peaking at about 1.48 million tons in 1968, due mainly to distant-water fleet

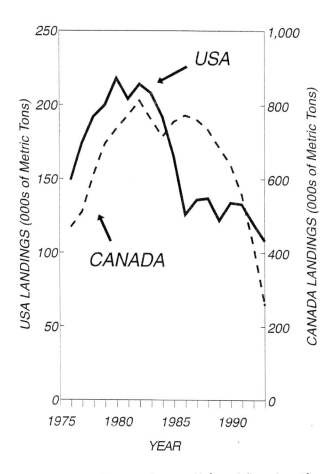

FIGURE 2.4.—Total landings of principal groundfish and flounders (thousands of tons) by Canada and the United States, 1976–1993. The vertical scale differs between the two graphs.

catches in subareas 2 (Labrador) and 3 (Newfoundland; Figure 2.6); catches then plummeted. After the United States and Canada extended their jurisdictions in 1977, cod landings again rose to around 600,000 tons in the mid-1980s, but declined sharply in the 1990s.

Of the 12 cod stocks, 8 are considered to be at or near historic low sizes and 3 are low to very low in abundance. Only the 3M (Flemish Cap) stock is at moderate size. Prospects for recruitment of new year-classes to the fishable populations are poor to very poor for all but the 3M and 4X stocks. Numerous cod stocks are now closed to directed fishing.

Labrador and Newfoundland

The Labrador 2GH stock produced catches averaging 5,000 tons during the century's middle years but exceeded 80,000 tons for several years in the 1960s, when about 99% was taken by distant-water fleets. Catches declined rapidly

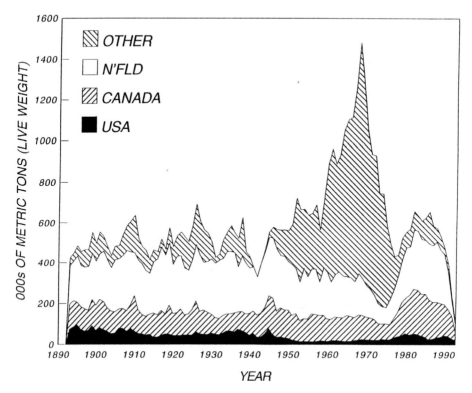

FIGURE 2.5.—Landings of Atlantic cod (thousands of tons), by country 1893–1993. Newfoundland retained dominion status until 1949, and its landings are separated from the rest of Canada's. "Other" denotes landings by non-North American fleets.

throughout the 1970s and dropped below 500 tons in 1985. No catches have been reported since 1990 apart from a few cod taken as bycatch in other fisheries (6). The stock shows no sign of recovery.

Until recently, the 2J+3KL cod stock off southern Labrador and Newfoundland (including the northern Grand Bank) yielded substantially more fish than the rest of the northern stocks combined and more than any cod stock in the western Atlantic. Annual catches were between 200,000 and 300,000 tons before 1960, but distant-water fishing pushed harvests to a peak of 800,000 tons in 1968. As elsewhere, catches then declined rapidly, recovered slightly in the latter 1980s, then continued to decline in parallel with estimated stock biomass (Figure 2.7). In July 1992, a moratorium was placed on commercial exploitation of the stock, and catches for personal consumption have since been reduced to about 1,000 tons per year. Associated with the collapse have been a contraction of the stock's geographic range, a decline in physical condition of the fish, and lower rates of growth and maturation. The stock size has continued its decrease during the moratorium, and it is very low. Projections of recovery time are not possible until there is evidence of substantial recruitment.

The cod fishery in division 3M (Flemish Cap) has traditionally been pursued by Portuguese trawlers and gillnetters, Spanish pair-trawlers, and Faroese longliners. Catches ranged from 22,000 to 33,000 tons during 1974–1979, fluctuated

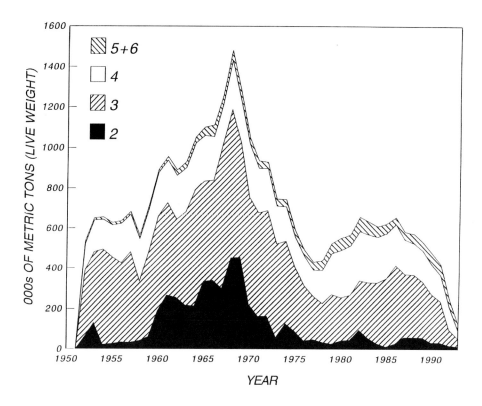

FIGURE 2.6.—Landings of Atlantic cod (thousands of tons) in NAFO subareas 2–6, 1952–1993.

around 11,000 tons between 1980 and 1987, declined to 2,000 tons in 1988, and increased to 40,000 tons in 1989; they have fluctuated between 11,000 and 32,000 tons since. The estimated 1994 catch from the 3M stock was 30,000 tons, the most from any cod stock in the western Atlantic. The fishable stock in 1994 was mainly composed of 3- and 4-year-old fish. Rebuilding of the spawning stock could occur if the exploitation rate of these cohorts were lowered in the next several years.

The southern Grand Bank (3NO) cod stock produced average landings of 64,000 tons between 1959 and 1994, peaking at 230,000 tons in 1967. Landings declined substantially in the 1990s as the stock declined to the lowest ever observed; 1994 landings were only 3,000 tons. Recruitment prospects remain poor for the stock, and thus recovery is uncertain.

The cod stock in subdivision 3Ps, off southern Newfoundland, supported average landings of 58,000 tons between 1959 and 1976 and 40,000 tons between 1977 and 1990. Landings have declined rapidly in the 1990s, and the fishery was closed by the Canadian government in August 1993. Fish stock abundance increased throughout the 1970s and early 1980s, peaking in 1988, then decreased rapidly. The 1995 research survey suggested that the stock has increased, but the survey index was dominated by a single very large catch and there was no evidence of notable recruitment. The stock is likely closer in size to that estimated during 1992–1994.

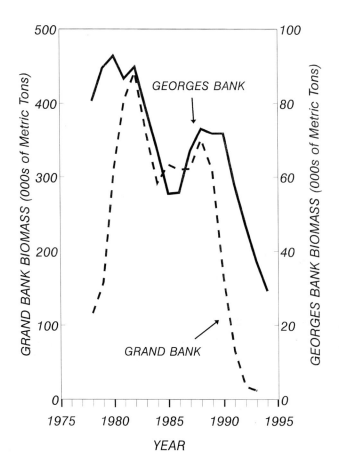

FIGURE 2.7.—Biomass (thousands of tons) of the 2J+3KL (northern Grand Bank) and 5Z+6 (Georges Bank) stocks of Atlantic cod, 1975–1995. Biomass estimates are for age-7 and older fish on Grand Bank (Bishop et al. 1994) and for spawning fish on Georges Bank (Serchuk et al. 1994b). Vertical scales differ between the two graphs.

Gulf of St. Lawrence

The two cod stocks in the Gulf of St. Lawrence, 3Pn+4RS to the north and 4TVn to the south, are separated by the Laurentian Channel. (Landings from 4Vn in November to April are assumed to belong to the southern Gulf of St. Lawrence stock.) Both stocks migrate out of the gulf in winter, the northern stock to 3Ps and the southern stock to 4Vn. The 3Pn+4RS stock of cod yielded an average of 72,000 tons from 1961 to 1994, and the 4TVn stock yielded 54,000 tons from 1950 to 1994. This harvest ratio of 0.75:1 for southern to northern gulf cod has been maintained historically. Southern Gulf of St. Lawrence cod share with the Labrador 2J fish the lowest weights at a given age of all the northwest Atlantic cod stocks. Both northern and southern gulf stocks increased rapidly after Canada extended its jurisdiction and remained relatively stable in the first half of the 1980s, but both have declined substantially since. Fisheries on both stocks are now closed, although 400 tons of northern gulf cod and 1,200 tons of southern gulf cod were landed in 1994 as bycatches from other fisheries. In the northern Gulf, strong indications are that natural mortality of adult fish may have increased in 1993–1994.

Scotian Shelf

The Scotian Shelf, extending from Cape Breton southwest along the Nova Scotian peninsula, has two cod stock complexes (4VsW and 4X) and one local stock (4Vn, harvested from May to October). The 4VsW and 4Vn cod fisheries are presently closed, but fishing for 4X fish is allowed. In 1994, 100 bycatch tons of 4Vn cod, 400 bycatch tons of 4VsW fish, and 13,000 tons of 4X cod were landed.

Adult biomass of 4VsW cod (combined weight of all adults in the stock) increased from about 25,000 tons in 1977 to 100,000 tons in 1986. It subsequently decreased to its lowest value ever in 1993 but may have increased slightly in 1994. The abundance of grey seals has increased substantially in the 4VW area, and consumption of small cod of ages 1 to 3 by these mammals could be high; if it is, this predation might be hampering recovery of the resource. The adult biomass of 4X cod has fluctuated less than in other stocks. It remained between 66,000 and 70,000 tons from 1971 to 1976 and has reached two peaks since then: 82,000 tons in 1980 and slightly above 60,000 tons in 1990. The lowest biomass ever recorded was slightly less than 30,000 tons in 1994. Biomass estimates are not calculated for 4Vn cod, but landings decreased from about 10,300 tons in 1986 to 2,300 tons in 1992, before stringent management measures were applied.

Gulf of Maine and Georges Bank

During the past 30 years, U.S. commercial landings from the Gulf of Maine stock (5Y) have ranged between 3,000 and 18,000 tons. Annual landings during 1960–1975 never exceeded 9,000 tons and averaged 5,500 tons; those during 1976–1985 were never less than 10,000 tons and averaged 12,000 tons. A record high catch of 18,000 tons occurred in 1991, but harvests dropped back to 8,000 tons in 1994. Fishing mortality in 1993 was nearly three times the management target established for this stock. In 1996, the fishery was still exploiting the remains of the 1987 year-class, which was by far the largest since 1980, but recent year-classes have been among the weakest on record. The Gulf of Maine cod stock is markedly depressed and remains overexploited.

Georges Bank is shared by Canada and the United States, and so is its cod stock (5Z+6). The Georges Bank cod management unit recognized by Canada occurs in statistical reporting areas that roughly align with the maritime boundary between the countries. Some U.S. landings are counted in this management area because statistical reporting areas slightly overlap the international boundary. Managers in the United States assume that cod on the Northeast Peak of Georges Bank are part of a larger stock that overlaps division 5Z and subarea 6. The following description is based on the larger U.S. unit, but the same trends apply to the smaller stock recognized by Canada. Since 1960, total commercial landings of Georges Bank cod have ranged between 11,000 (1960) and 57,000 tons (1982). Landings by U.S. vessels increased fourfold between 1960 and 1980 (10,800 to 40,000 tons) but declined to 18,000 tons in 1986, ranged from 24,000 to 28,000 tons between 1988 and 1991, and declined sharply to 14,600 in 1993 and 9,800 tons in 1994; a similar trend is shown by estimated stock biomass (Figure 2.7). Canadian catches peaked in 1990 at 14,300 tons, but have since declined steadily to 5,300 tons in 1994 and much less thereafter. The 1990 year-class currently dominates the stock; older fish are almost nonexistent and incoming year-classes are relatively weak. The Georges Bank cod stock is very small relative to historical levels.

Haddock

Landings of haddock were low in the 1800s because the species did not keep well when salted and consumption of fresh fish was localized in coastal areas. The U.S. haddock "industry" began in the early 1900s, after the first steam-driven otter trawler was introduced in 1906 (7). Haddock landings expanded rapidly as resources on Georges Bank and the Scotian Shelf became increasingly targeted by trawl fleets in both the United States and Canada; comprehensive landings statistics for haddock began in 1910 for Canada (Figures 2.8, 2.9). A peak in landings in the early 1930s was followed by a rapid decline, primarily because stock sizes became smaller on Georges Bank (8). After the early 1930s, the U.S. fishery expanded onto the Scotian Shelf, and total landings by U.S. fishers fluctuated around 50,000 tons between the mid-1930s and the 1960s. Total landings in the NAFO area built up rapidly in the 1950s and 1960s as a result of distant-water fleet catches (Figure 2.8). Landings peaked in 1965 at 249,000 tons (primarily due to exploitation of the exceptional 1963 year-classes on Georges and Browns banks), but declined to only 23,000 tons by 1974. Catches temporarily improved in the early 1980s, but then declined to an historic low of 14,000 tons in 1993.

The U.S. proportion of haddock landings has declined steadily from about 50% of the catch in 1950 to less than 10% in 1993. Subareas 4 (Scotian Shelf) and 5 (Georges Bank–Gulf of Maine) accounted for most of the haddock catch except for some strong landings from the Grand Bank and St. Pierre Bank (subarea 3) in the 1950s and early 1960s (Figure 2.9).

Newfoundland

Fisheries for haddock stocks in the Newfoundland area—Grand Bank (3LNO) and St. Pierre Bank (3Ps)—developed in the late 1940s and 1950s. Landings of Grand Bank haddock increased from 8,500 tons in 1953 to 76,000 tons in 1961, then declined very quickly to 1,700 tons in 1969 and to less than 1,000 tons from 1976 to 1983. The 1980 and 1981 year-classes were strong but were largely harvested before they reached reproductive age. Exploitation of St. Pierre Bank haddock was similar: 5,800 tons in 1953, 58,000 tons in 1955, and 6,000 tons in 1957. The 1981 year-class was strong on St. Pierre Bank as well as on Grand Bank, but it was heavily exploited before it matured. Although strong year-classes were produced in both stocks during the 1980s, the fishing pressure was so high that the strong recruitment was not allowed to rebuild the stocks; too much of the catch consisted of juveniles.

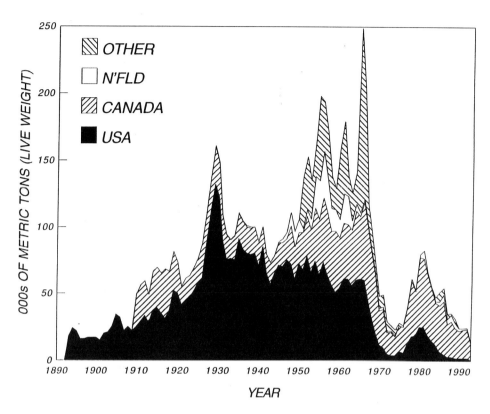

FIGURE 2.8.—Landings of haddock (thousands of tons) by country, 1893–1993. "Other" denotes landings by non-North American fleets.

Scotian Shelf

The Scotian Shelf has an eastern 4TVW stock of haddock and a western 4X stock that extends into the Bay of Fundy. Haddock have been scarce in the Gulf of St. Lawrence (4T) for several years, and the Gulf population may draw on the eastern Scotian Shelf stock when the latter is abundant. The eastern Scotian Shelf stock was heavily exploited from the late 1940s to the late 1960s; from a peak of 55,500 tons in 1965, landings dropped to 24,000 tons in 1966 and then declined steadily to 1,400 tons in 1976. After Canada extended its jurisdiction, harvests rebounded to 20,000 tons in 1981, but they were generally below 10,000 tons after 1982 and less than 100 tons in 1994. Haddock on the western Scotian Shelf and in the Bay of Fundy also were objects of distant-water fishing, which targeted the extraordinarily strong 1963 year-class. Landings peaked at 42,000 tons in 1966 and declined to 13,000 tons in 1973. Landings then increased to 31,000 tons in 1981, fell to 6,700 tons in 1989, rose to 10,000 tons in 1992, and dropped to 4,300 tons in 1994.

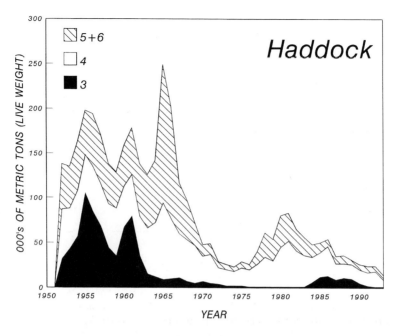

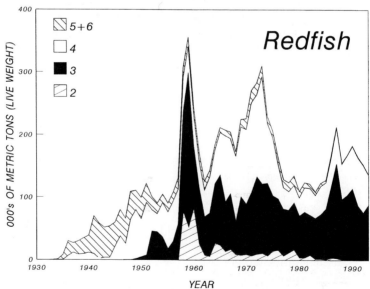

FIGURE 2.9.—Landings of haddock and redfish (thousands of tons) in NAFO subareas 2–6. Haddock landings are given for 1952–1993, whereas redfish landings are given for 1935–1993.

Gulf of Maine and Georges Bank

Landings of Gulf of Maine haddock in the United States (division 5Y) declined from about 5,000 tons annually in the early 1960s to less than 1,000 tons in

1972. Landings increased sharply between 1974 and 1980, reaching 7,300 tons in 1980. Since 1983, catches have declined to record lows (500 tons or less since 1988). Virtually all landings from this stock are now taken in the U.S. fishery. Abundance remains near its all-time low and recruitment has been insufficient to maintain landings near their historical average.

For management and assessment purposes, Canada and the United States use different stock units for Georges Bank haddock, as for Atlantic cod. The Canadian haddock unit comprises only the eastern-most portion of Georges Bank (5Z) whereas the U.S. stock unit comprises all of division 5Z, including that used by Canada, and division 6. The following description is based on the larger U.S. unit, but the same trends generally apply to the Canadian unit.

Landings of Georges Bank haddock increased from about 50,000 tons annually prior to 1965 to nearly triple that amount in 1965 and 1966, due to intense fishing by the distant-water fleets on the extraordinarily strong 1962 and 1963 year-classes. Landings declined sharply to 4,300 tons in 1976 but increased between 1977 and 1980, reaching about 28,000 tons. Landings declined after 1980, dropping to 4,500 tons in 1989 and have since ranged between 4,000 and 7,000 tons. In 1993, landings were 4,400 tons, the second lowest ever. Landings by U.S. fishers accounted for 16% (700 tons) of the 1993 total; Canadian fishers provided the difference. Recruitment was poor during most of the 1980s. The 1989 and 1990 year-classes continued this trend, each producing only about 4 million fish at age 1. The 1992 year-class is the largest since 1987, and it has caused an approximate doubling of stock biomass to over 20,000 tons. Year-classes spawned after 1992 have not been as strong. Fishing mortality of this stock has been reduced by low Canadian quotas and increased U.S. protection.

Redfish

Before the mid-1930s, the northwestern Atlantic redfish fishery provided a few fish for local consumption, but it was not organized as an industry until the mid-1930s (9). The U.S. fishery was first to develop, supplying "ocean perch" to midwestern markets as a substitute for yellow perch from the Great Lakes. Catches in the 1930s and 1940s were primarily from the Gulf of Maine (subarea 5), but they expanded rapidly eastward to the Scotian Shelf (subarea 4) and Newfoundland (subarea 3) in the 1950s as local stocks were depleted. Substantial catches off Labrador (subarea 2) in the late 1950s and 1960s have declined to a very low percentage of the total (Figure 2.9). Total landings peaked in 1959 at 356,000 tons (Figure 2.10), primarily as a result of large catches by the former USSR off Labrador (subarea 2) and Newfoundland (subarea 3). Another peak occurred in 1973 at 310,000 tons, due primarily to Canadian landings (most of which were taken in the Gulf of St. Lawrence). The current U.S. fishery for redfish is virtually nonexistent, though some redfish are landed as bycatch in fisheries directed to other stocks. Most landings are now taken from Newfoundland and the Scotian Shelf; about half of the landings are made by Canada and half by other NAFO countries (Figure 2.10).

Before 1993, redfish in the Gulf of St. Lawrence, in southern Newfoundland, and on the Scotian Shelf were managed as three units in divisions 3P, 4RST, and 4VWX. Late-summer surveys in divisions 3P and 4V showed a continuous distribution of redfish across the division line. Three new management units were pro-

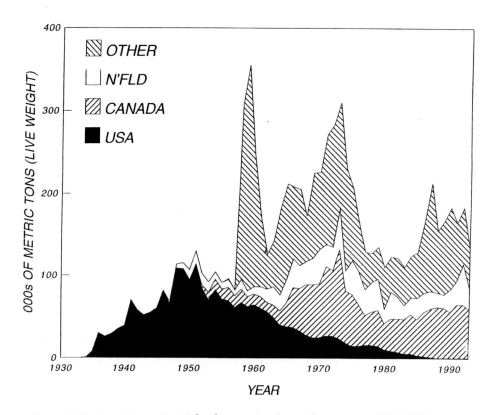

FIGURE 2.10.—Landings of redfish (thousands of tons) by country, 1935–1993.

posed by Canadian scientists and implemented in 1993: unit 1 redfish are those in the Gulf of St. Lawrence plus those caught in 3Pn or in 4Vn during January–June, unit 2 comprises redfish in subdivisions 3Ps+4Vs and 4W (part) for the whole year and in 3Pn+4Vn for July–December, and unit 3 comprises redfish in the rest of divisions 4W and in 4X. In the following discussion, unit 1 redfish are treated as being in the Gulf of St. Lawrence area, unit 2 redfish in the Newfoundland area, and unit 3 redfish in the Scotian Shelf area.

Redfish catches may contain three species: Acadian, deepwater, and golden redfish. Except on Flemish Cap, golden redfish are not abundant and play an insignificant role in commercial fisheries. The expected species mix of the "beaked" (Acadian and deepwater) redfishes varies along a north-to-south gradient: deepwater redfish predominate in northern waters (2+3K) and Acadian redfish in more southern waters (unit 3). In the Gulf of St. Lawrence, deepwater redfish predominate despite the low latitude. In U.S. waters, Acadian redfish are dominant.

Newfoundland

Five redfish stocks are exploited in the Newfoundland area: 2+3K, 3M, 3LN, 3O, and unit 2 (see above). The 3LN, 3M, and 3O stocks traditionally have been and continue to be exploited mainly by foreign countries. The 2+3K redfish yielded average landings of 27,000 tons between 1961 and 1986, but landings declined from 18,500 tons in 1986 to 280 tons in 1991 because of continuously poor recruitment since 1971. Landings of Flemish Cap (3M) redfish have been at or above 20,000 tons since 1983, reaching a maximum of 81,000 tons in 1990 and declining to 29,000 tons in 1993. Landings of 3LN redfish peaked at almost 80,000 tons in 1987 but declined to 24,000 tons in 1993. Surveys in division 3O showed an increase in biomass from about 10,000 tons in the spring of 1992 to 84,000 tons in the spring of 1995. Because most of the 3O stock consists of small redfish, emigration from 3LN or unit 2 is a possible explanation for the apparent increase in biomass, although indigenous recruitment cannot be discounted. Redfish in the trawlable areas of division 3O are small and until recently were not fished by the Canadian industry. Russia has historically dominated this fishery. Landings declined from 16,000 tons in 1993 to 4,000 tons in 1994.

Gulf of St. Lawrence

The only redfish stock in the Gulf of St. Lawrence is unit 1. It has two species, which are caught in 4RST and (during January–June) in 3Pn and 4Vn. The fishery started in the early 1950s and expanded rapidly to almost 50,000 tons in 1955, but it subsequently declined to less than 10,000 tons by 1962. Landings increased to 95,000 tons in 1968 and remained between 80,000 and 90,000 tons until 1973, when landings increased abruptly to 136,000 tons. Landings then declined rapidly to 15,000 tons in 1978 and increased slowly afterwards, reaching about 77,000 tons in 1992. Surveys have shown a rapid biomass decline, and the fishery was closed in 1995.

Scotian Shelf

Redfish on the Scotian Shelf (unit 3 in 4W[part]+X) have been subjected to low exploitation since extension of jurisdiction. Landings increased to about 5,200 tons in 1993 and 1994.

Gulf of Maine

During development of the Gulf of Maine redfish fishery, U.S. landings rose rapidly to a peak at about 60,000 tons in 1942, followed by a gradual decline to less than 10,000 tons during the mid-1960s. They ranged from 10,000 to 16,000 tons during the 1970s but declined continuously throughout the 1980s, reaching historical lows between 500 and 600 tons per year during 1989–1991. Landings in 1992 and 1993 increased slightly to 800 tons. Stock biomass has increased over levels of the early 1980s, though it remains well below the average of the 1960s and early 1970s. The biomass increase in 1990–1993 reflects accumulated recruitment and growth of one or more recent year-classes. Recruitment was poor except for moderate year-classes in 1978 and the mid-1980s, though it has improved somewhat in the 1990s. The population lacks older age-groups and remains overexploited.

Silver Hake

The silver hake resource of the region comprises three stock units: Scotian Shelf (4VWX), Gulf of Maine northern Georges Bank (5YZe[part]), and southern Georges Bank–Middle Atlantic Bight (5WZ[part]+6). The Scotian Shelf and southern Georges Bank–Middle Atlantic stocks were historically dominant. They supported very high catches for several years when the distant-water fleets were active in the region.

Scotian Shelf

Silver hake has been the dominant species taken in distant-water fisheries on the Scotian Shelf since the early 1960s; the former USSR and Cuba were the best-represented foreign countries (10). A Canadian domestic fishery has harvested silver hake since 1990, primarily through charter arrangements with Russian and Cuban fishing vessels. A small domestic fishery also is developing in the LaHave and Emerald basins. Total catches have decreased from 70,000 tons in 1990 to 8,000 tons in 1994.

Gulf of Maine and Northern Georges Bank

Following the arrival of distant-water fleets in 1962, total landings of silver hake from the Gulf of Maine and northern Georges Bank increased rapidly to a peak of 95,000 tons in 1964 but then declined for 13 years to 3,000 tons in 1979. Before 1976, distant-water catches averaged about 49% of total. Activity by distant-water fleets diminished after 1977 and domestic landings have remained less than 10,000 tons per year since 1979. Commercial landings in 1993 were 4,000 tons, the lowest since 1981. Biomass indices have increased significantly (with fluctuation), and recent recruitment appears to be at or above those of the mid-1970s.

Southern Georges Bank and Middle Atlantic Bight

From 1955 to 1965, U.S. landings of silver hake from southern Georges Bank and the Middle Atlantic Bight averaged 15,000 tons per year. Foreign landings were much greater, reaching more than 280,000 tons in 1965. From 1966 to 1971, both U.S. landings (averaging 8,000 tons) and foreign landings (averaging 80,000 tons) declined. After 1971, domestic landings steadily increased from a low of 5,000 to 11,000 tons in 1978, a level that has been maintained since. Distant-water fleets caught more than 40,000 tons annually from 1971 to 1977, but they have taken negligible amounts since 1982. Domestic U.S. landings varied between 10,000 and 14,000 tons averaging 12,000 tons, during 1989–1993. Stock abundance is low and continues to decline. The stock is overexploited and will continue to be so until mortality rates of juveniles are minimized and overall fishing mortality is markedly reduced. Recently, a "juvenile whiting" fishery has developed in response to an export market for small silver hake that traditionally have been discarded.

Pollock

The pollock resource in the region comprises at least two stock units, one on St. Pierre Bank (3Ps) and one (perhaps more) in the Scotian Shelf–Gulf of Maine region (4VWX+5). The southern stock has been the more important one; average catches at St. Pierre Bank averaged only 1,000 tons between 1967 and 1994.

Scotian Shelf and Gulf of Maine

For management and assessment, Canada and the United States use different stock units for the pollock resource in the Scotian Shelf–Gulf of Maine area. The Canadian stock unit includes the western Scotian Shelf and eastern Georges Bank, whereas the United States treats Scotian Shelf, Georges Bank, and Gulf of Maine pollock as a single unit. The following description is based on the larger unit, but trends generally apply to the smaller Canadian one as well.

Canadian pollock landings increased steadily from 25,000 tons in 1977 to an average of 44,000 tons during 1985–1987. Landings declined gradually through 1991, and then dropped sharply in 1992 and 1993. The total U.S. and Canadian catch from the stock was 47,400 tons in 1991 and 27,000 tons in 1993. The decline in combined landings since 1986 was led by the sharp U.S. reductions during 1987–1989, followed by the Canadian declines in 1992–1993. Landings paralleled total stock size, which increased through the late 1970s and early 1980s and then declined. The last strong pollock year-class was produced in 1988.

Landings in the United States increased from an average of 10,000 tons during 1973–1977 to more than 18,000 tons annually between 1984 and 1987, peaking at 24,500 tons in 1986. Since then, annual U.S. landings have steadily declined to 6,000 tons in 1993, a 77% decrease from 1986. These U.S. catches have generally accounted for 20 to 30% of the total harvest from this stock. Since 1984, U.S. fishing has been restricted to that part of the stock occurring west of the international boundary through the Gulf of Maine and Georges Bank.

The record high landings during the mid-1980s (exceeding 63,000 tons per year between 1985 and 1987) stimulated relatively high fishing mortality during the rest of the decade. Although pollock statistics in both countries may include some catches of cod and haddock that were misreported when quotas for those species were exceeded, the pollock stock is considered to be at least fully exploited.

American Plaice

The American plaice resource in the northwestern Atlantic comprises six defined stocks. The southern Grand Bank stock (3LNO) was historically the most productive, averaging nearly twice the production of all the others combined. All plaice stocks are considered to be in poor to moderate condition. Recent yields are among the lowest recorded. Scotian Shelf area plaice (4VW and 4X) are considered part of the "flatfish complex."

Newfoundland

The Newfoundland area has four American plaice stocks: 2+3K, 3LNO, 3M, and 3Ps. Historically, 3LNO has supported the largest plaice fishery in the northwest Atlantic; peak landings were almost 95,000 tons in 1967. Landings were reasonably stable, mostly between 40,000 and 50,000 tons, between 1974 and 1985. They increased to near 65,000 tons in 1986, then steadily declined to 7,000 tons in 1994. Large numbers of small plaice were harvested after 1986, presumably accelerating the decline. There was no directed fishing in 1994 and a moratorium has been in effect thereafter. The second largest plaice fishery was prosecuted in subdivision 3Ps. Landings peaked there at almost 15,000 tons in 1973 but have decreased to less than 5,000 tons since 1983; only 800 tons were taken in 1993 and 100 tons in 1994. Landings of 2+3K plaice increased steadily during the 1960s to almost 13,000 tons in 1970, but they have rarely exceeded 2,000 tons since 1980 and biomass now is very low. Plaice on the Flemish Cap (3M) are mainly taken as bycatch in the cod and Greenland halibut fisheries. Landings peaked at 5,600 tons in 1987 but declined to 700 tons in 1994.

Gulf of St. Lawrence

One American plaice stock exists in the Gulf of St. Lawrence (division 4T). The fishery was exploited mainly by longlines in the 1930s, became predominantly an otter trawl and seine fishery from the late 1940s to 1960s, and diversified somewhat thereafter (11). Once considered a bycatch of the directed cod fishery, plaice have increasingly been targeted in the Gulf of St. Lawrence. Landings declined sharply from the mid-1980s, when they averaged about 10,000 tons. Discarding of undersized plaice has been a consistent problem in this fishery despite management measures. Surveys indicate biomass is currently at its lowest level of the past 24 years.

Gulf of Maine and Georges Bank

Landings of Gulf of Maine–Georges Bank (5YZ) American plaice increased from an average of 2,000 tons during 1972–1976 to 13,000 tons per year during 1979–1984, then declined; since 1985, they have ranged between 2,300 and 7,000 tons (6,600 tons in 1992 and 5,800 tons in 1993). Abundance and biomass indices reached record-low values in 1987, but they increased through 1990 as the strong 1987 year-class recruited to the fishery. These indices declined in 1991 and 1992 but markedly increased in 1993 due to improved recruitment from the 1989 and 1990 year-classes. The stock is considered to be at a moderate level of abundance, but remains overexploited.

Witch Flounder

Witch flounder resources of the region comprise seven stocks (the 4VW and 4X stocks are considered part of the "flatfish complex"). Historically, the resources off Labrador, the eastern Scotian Shelf, and the eastern Grand Bank provided the greatest yields, but most catch now comes from the Gulf of Maine and Georges Bank region.

Canada

Canada has six witch flounder stocks: 2J+3KL, 3NO, 3Ps, 4RST, 4VW, and 4X. The largest catches, in decreasing order, have been made off Labrador and the northern Grand Bank (2J+3KL; 24,000 tons in 1973), eastern Scotian Shelf (4VW; 21,000 tons in 1968), southern Grand Bank (3NO; 15,000 tons in 1971), Gulf of St. Lawrence (4RST; 7,000 tons in 1976), St. Pierre Bank (3Ps; 4,800 tons 1967), and western Scotian Shelf–Bay of Fundy (4X; 1,100 tons in 1971). For all stocks combined, the landings peaked at 55,000 tons in 1971 and averaged 22,000 tons since 1960. Since the 1970s, witch flounder stocks have declined drastically, and landings have declined accordingly. Witch flounder biomass in 4VW remains near the lowest observed. In 4X, the 1994 biomass index was the lowest observed, only about one-third the maximum observed in 1978.

Gulf of Maine and Georges Bank

Predominance of witch flounder in U.S. landings has alternated between Georges Bank and the Gulf of Maine. In recent years, most of the U.S. catch has come from the Gulf of Maine. Canadian catches from both areas have been minor (never more than 68 tons annually). Distant-water fleet catches in U.S. waters averaged 2,700 tons during 1971–1972, but subsequently declined sharply and have been negligible since 1976. After averaging 2,800 tons during 1973–1981, total landings increased sharply during the early 1980s and peaked at 6,500 tons in 1984. Landings then steadily declined through 1990, reaching 1,500 tons, the lowest value since 1964. Landings have since increased slightly to 2,600 tons in 1993. Stock biomass sharply declined from 26,000 tons in 1982 to about 6,300 tons in 1990 and has fluctuated at about 7,000 tons through 1993. Due to continued growth and maturation of the strong 1990 year-class, witch flounder biomass is expected to increase in the short term (through 1996), but will thereafter decline unless fishing mortality is reduced. The stock biomass is low and the stock is overexploited.

Yellowtail Flounder

The yellowtail flounder resource consists of five stocks and two stock complexes. The dominant resources are those on Grand Bank, in southern New England, and on Georges Bank (12).

Newfoundland

Landings of Grand Bank yellowtail flounder (subdivision 3LNO) reached 39,000 tons in 1972, declined to 8,000 tons in 1976, increased again to 30,000 tons in 1986, but then declined to 2,000 tons in 1994. The stock biomass is low, and the stock growth that could have resulted from the strong 1984–1986 year-classes has not occurred, presumably because high exploitation reduced those year-classes at a young age. The stock area has contracted in recent years, making the stock more vulnerable to overexploitation. No directed fishery was allowed in 1995.

Georges Bank

Total landings of yellowtail flounder from Georges Bank (5Ze) averaged 16,000 tons during 1962–1976 but declined to an average of 6,000 tons between 1978 and 1981. Strong recruitment from the 1977 and 1980 year-classes and high fishing effort allowed landings to exceed 10,500 tons in 1982 and 1983. Since 1985, landings have been 3,000 tons or less. Landings fell to a record low of 1,100 tons in 1989, increased to 2,900 tons in 1992, but declined to 2,300 tons in 1993. Current biomass on Georges Bank is about 10% of biomasses estimated in the 1960s. The average fishing mortality rate during 1983–1993 was twice the overfishing criterion for this stock. The stock is overexploited, abundance is low, and few age-groups are present.

Southern New England

Total landings of yellowtail flounder from the southern New England stock (5Zx) averaged 20,000 tons during 1963–1968, but declined abruptly after 33,000 tons were taken in 1969. Landings fell by 75% to 9,000 tons in 1971 and have exceeded that value only three times in the past 24 years. A short-lived recovery occurred in 1990 when the strong 1987 year-class became fully recruited. The apparent recoveries in 1983 and 1990 have produced landings roughly one-half of the preceding maximum values. In 1993, landings totaled only 500 tons, a record low. Current stock size is about 5% of values observed in the late 1960s. This stock is overexploited and stock size is extremely low. Fishing mortality during 1990–1992 was four times the overfishing criterion for the stock.

Cape Cod

Traditionally, landings of Cape Cod yellowtail flounder (5YZw[part]) have been much smaller than landings from the southern New England and Georges Bank stocks. In 1993, for the first time, landings from the Cape Cod stock (800 tons) exceeded those from southern New England. Landings of Cape Cod yellowtail flounder fluctuated between 1,000 and 2,000 tons in the 1960s, increased during the 1970s, attained a record-high 5,000 tons in 1980, then declined to a record low of 800 tons in 1992 and 1993. Recent declines in landings and relatively low abundance indices (compared to those in the mid-1970s) suggest a reduction in biomass. The stock is considered to be overexploited.

Greenland Halibut

Two stocks of Greenland halibut, also called "turbot," are believed to exist in the northwest Atlantic. The smaller one is in the Gulf of St. Lawrence; the considerably larger one extends from Baffin Island and Greenland (subareas 0 and 1) past Labrador (subarea 2) to the southern tip of Grand Bank in subarea 3.

Landings of Greenland halibut from subareas 2 and 3 increased sharply in 1990 as a new fishery developed beyond Canadian jurisdiction, and they remained high in 1992–1994. The fishery is not well documented, but NAFO data (including those from subareas 0 and 1) indicate that the average age of fish in the catch and the catch per unit of fishing effort both declined between the late 1980s and 1994, suggesting a reduction in stock size.

Greenland halibut landings in the Gulf of St. Lawrence peaked at 8,800 tons in 1979 and at 11,000 tons in 1987 before decreasing to 2,300 tons in 1991. They were about 3,000 tons in 1992–1993, and probably exceeded 5,000 tons in 1994. Fish longer than 50 cm made up half of the harvest in the 1970s and early 1980s, but such large fish now are rare.

THE INDUSTRY

Canada

At the turn of the Twentieth Century, the Canadian groundfish industry was dominated by production of salt cod in Newfoundland (13). Fishing gears were primarily coastal traps and baited lines, and annual production of cod averaged about 200,000 tons. Traditional fishing methods changed little for the first two decades of this century. The Canadian government restricted development of a trawl fishery in the 1920s and 1930s because traditional fishers opposed it. The fishery prospered during World War I and generally declined during the Great Depression. After World War II, trawling became more popular, and trawl fleets developed in all regions; the first trawler was authorized to fish in Canada in 1947. The number of large Canadian fishing vessels (50 gross registered tons or more) increased from 211 in 1959 to 558 in 1968 in concert with the buildup of the distant-water fleet (Figure 2.11).

After Canada extended its fisheries jurisdiction in 1977, Canadian fishing effort expanded rapidly, particularly with respect to offshore trawling from Newfoundland and nearshore trawling and long-lining on the Scotian Shelf and in the Bay of Fundy (Figures 2.11–2.13). New export strategies favored the production of fresh and frozen groundfish over salt fish, which allowed greater access to U.S. and European Community markets. The vast majority of Canadian groundfish production in the 1980s and 1990s supplied those countries as well as markets in the Caribbean and Latin America.

Recent declines in fish stocks have been accompanied by major reductions in fishing effort and, correspondingly, in harvesting and processing employment. Canadian fishing effort off Newfoundland was reduced from about 200,000 vessel-hours per year in 1991 to about 25,000 hours in 1994 (Figure 2.12). The closure of cod fisheries in the Gulf of St. Lawrence resulted in major effort reductions there as effort was directed to a few minor species. In the Scotian Shelf–Bay of Fundy region, effort by offshore trawlers has steadily declined since extention of jurisdiction, whereas inshore trawler effort rose by a factor of about six. Inshore trawler effort peaked in 1991 (Figure 2.13) before declining by about half by 1994. Long-line effort in the Scotian Shelf–Bay of Fundy region also generally increased until those fisheries were closed (Figure 2.13).

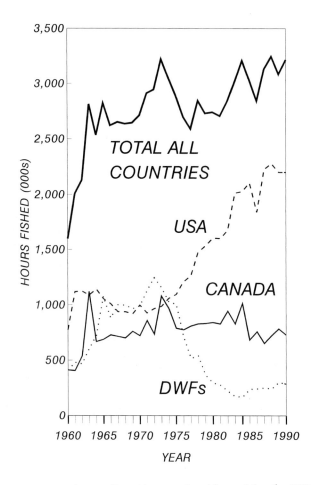

FIGURE 2.11.—Reported fishing effort (thousands of hours) by the U.S. and Canadian vessels, distant-water fleets (DWFs), and all vessels working in the northwest Atlantic, as compiled by ICNAF and NAFO, 1960–1990. Data are for all fisheries combined and are not standardized to vessel size.

United States

Groundfish landings in U.S. waters are primarily derived from Georges Bank (5Ze), the Gulf of Maine (5Y), and southern New England (5Zw+6A). Offshore groundfish fisheries developed on Georges Bank between 1720 and 1750, primarily for Atlantic cod, but they expanded during the early 1800s to include Atlantic halibut and Atlantic mackerel (14). Technological improvements (e.g., purse seines, line trawls, otter trawls) and increased market demand for iced, fresh fish stimulated rapid growth and diversification of the Georges Bank fishery. Landings of Georges Bank haddock increased during the late 1870s and early 1880s, as did catches of hake, cusk, and pollock.

In the early part of the Twentieth Century, steam- and diesel-powered vessels and freezing and refrigeration technology transformed the character of the offshore fisheries. Coupled with increased use of otter trawls and power-driven

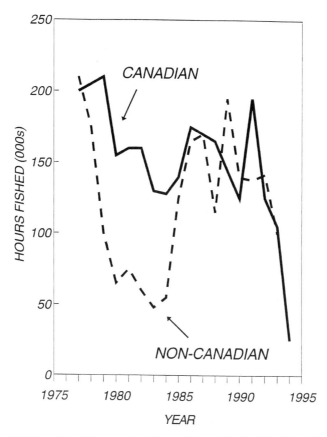

FIGURE 2.12.—Fishing effort (thousands of vessel hours) in the Newfoundland area, 1977–1994 (redrawn from Anonymous 1995c).

fishing equipment, these developments significantly enhanced the mobility, profitability, and collective fishing power of the Georges Bank fishing fleets. By the early 1920s, haddock had become the mainstay of the groundfish fishery, accounting for about two-thirds of the U.S. groundfish catch from the bank. Flounder catches also increased during the 1920s when proper filleting techniques were introduced and the winter and yellowtail flounder stocks on the southern flank of Georges Bank were found in commercial quantities.

Between 1930 and 1960, the U.S. groundfish fishery was relatively stable except for the intervention of World War II. In general, fishery growth did not exceed resource capacities, although concern was raised over the large quantities of small haddock discarded in the Georges Bank fishery.

An important redfish fishery developed in the Gulf of Maine beginning in the mid-1930s. New England vessels supplied new markets in the midwestern United States, and the volume of this fishery expanded rapidly. Initially the fishery was prosecuted in subarea 5, but it developed eastward into subareas 4 and 3. The U.S. fishery for redfish is all but gone now, due to the loss of fishing grounds off Canada and the overfishing of U.S. resources.

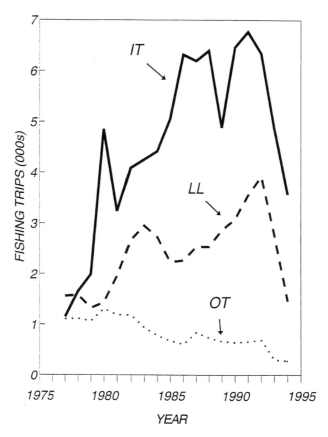

FIGURE 2.13.—Fishing effort (thousands of trips) to the Scotian Shelf and Bay of Fundy by inshore otter trawlers (IT), offshore otter trawlers (OT), and long-liners (LL), 1977–1994. Data are from Anonymous (1995d).

Before 1960, Georges Bank was fished almost exclusively by the United States. During the 1960s and early 1970s, however, distant-water fleets arrived from the former USSR, Poland, German Democratic Republic, Federal Republic of Germany, Japan, and other countries. Although these fleets initially exploited Atlantic herring, effort was soon directed towards abundant stocks of groundfish and flounders. Total yield from Georges Bank increased sharply from 240,000 tons in 1960 to about 780,000 tons in 1965 due to substantial increases in the groundfish component (principally haddock and silver hake). Subsequently, groundfish landings sharply declined and fishing effort was redirected to pelagic species, principally herring (later mackerel) and squids.

After the United States extended its jurisdiction, its groundfish fleet expanded rapidly. Total fishing effort by otter trawlers doubled from 1976 to 1985 and remained at a stable high level through 1993 (Figure 2.14). Catches and catch rates by the expanding U.S. fishery increased in the late 1970s, primarily due to good year-classes of haddock and cod (Figure 2.4). However, the fishing pressure gen-

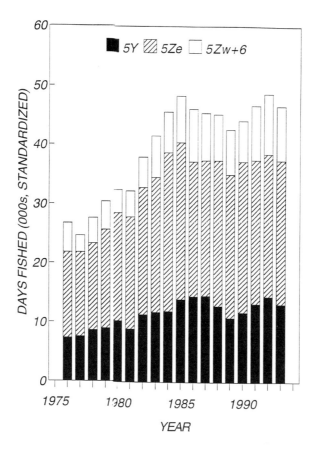

FIGURE 2.14.—Otter trawl fishing effort (thousands of days) off the northeastern United States in subareas 5 and 6, 1976–1993. Data are from NEFSC (1995). Effort has been adjusted ("standardized") for differences in catch rate by various vessel size-classes.

erated by the increased fleet size and fishing power increased rapidly. Important stocks of haddock, yellowtail flounder, and other species have collapsed, and declines in the Atlantic cod resource have been drastic.

More than 4,300 groundfish vessel permits were issued for waters off the northeastern United States. Because of reduced landings and increasingly restrictive management measures enacted to protect traditional species, the U.S. groundfish industry has increasingly targeted "other" groundfish species, including goosefish (monkfish), spiny dogfish, skates, and white hake. Landings of these species have expanded rapidly, both creating export opportunities (skates and dogfish), and replacing traditional resources in the U.S. market. However, these resources have finite productivity, and some of them may be overfished already (e.g., goosefish).

The United States remains the primary market for groundfish caught off eastern North America. Most domestic U.S. production is consumed internally; the majority of Canadian production is exported to the United States. Imports of Canadian cod and other groundfish into the United States expanded rapidly between 1978 and 1989 (Figure 2.15) accounting for up to two-thirds of Canadian cod production in some years. These imports declined substantially in the 1990s as cod production in Canada dropped. The strong U.S. market demand has led to greater imports of groundfish from other countries and substitution (both foreign and domestic) of other species.

Employment in both harvesting and processing sectors of the groundfish industry has declined substantially since the 1980s. Given the poor short-term prospects for resource recovery and the likelihood of even more restrictive measures in the United States, employment is likely to continue to decline, particularly in the harvesting sector. The U.S. government has instituted an effort to buy back groundfish permits and has committed funds for this purpose. Thus, it is likely that the number of active groundfish permits and effective fishing effort will decline in the next several years.

MANAGEMENT STRUCTURES

Despite relatively low distant-water catches in the northwest Atlantic at the time, an international management forum for the region was created in 1951. As a result of negotiations during the late 1940s, the International Commission for the Northwest Atlantic Fisheries (ICNAF) assumed the role of coordinating management of offshore fisheries in the region (15). As fisheries built up, ICNAF became the principal focus for fishery management beyond the (then) 22-km (12-mile) national jurisdictions. It instituted mesh size restrictions, reporting requirements, inspection on the high seas, and (in the early 1970s) quota management. Quota management substantially reduced the directed fishing on groundfish stocks by distant-water fleets, which had become a major problem in the 1960s. Nevertheless, the rapid decline of historically important stocks and the corresponding reductions in Canadian and U.S. landings brought growing domestic discontentment with the still-high fishing effort by distant-water fleets.

The United States and Canada extended their fishery jurisdictions in 1977, in large part to counteract the influences of distant-water fishing. The United States, upon asserting its control over the territorial sea out to 370 km (200 nautical miles), withdrew from ICNAF. That body was then reconstituted as the Northwest Atlantic Fisheries Organization (NAFO) with Bulgaria, Canada, Cuba, Denmark (for the Faroe Islands), the European Economic Community, German Democratic Republic, Iceland, Japan, Norway, Poland, Portugal, Romania, and the USSR as members. The United States maintained observer status in NAFO for a number of years and formally joined it in 1995. The organization currently coordinates the management of 11 principal groundfish and flounder resources in subareas 2–6 and additional resources in areas 0 and 1. It establishes annual quotas, allocations by country, and other technical measures as required.

Extended jurisdictions brought lingering disagreement between Canada and the United States over the position of their mutual Atlantic boundary. In October 1984, the International Court of Justice delimited a maritime boundary between the two countries across the Gulf of Maine and Georges Bank. This decision effec-

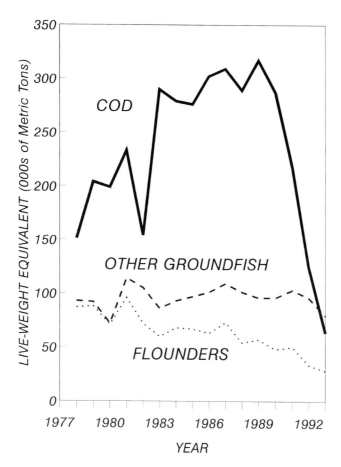

FIGURE 2.15.—Imports of Canadian-caught groundfish (thousands of tons) into New England, 1978–1993. Data are live-weight equivalents for cod, flounders, and other groundfish species (NEFSC 1995 and earlier documents in that series).

tively partitioned the management and use of many Georges Bank fish stocks (16). Management institutions developed in both countries since 1977 are described in the following sections.

Canada

After Canada extended its jurisdiction, it developed a logical evolution from the ICNAF process. It set up an independent scientific committee, the Canadian Atlantic Fisheries Scientific Advisory Committee, which corresponded to the ICNAF Standing Committee on Research and Statistics, and an independent board of fishery advisors, the Atlantic Groundfish Advisory Committee.

The Canadian Atlantic Scientific Advisory Committee (CAFSAC) was created to peer review scientific analyses and provide advice on all biological, technical, sociological, and economic matters for all species. In practice, the peer review and advisory functions became limited to biological and technical questions. Seven subcommittees were created, five of them dealing with species groups and two with "themes" (one on marine environments and ecosystems, the other on statistics, sampling, and surveys). Only government employees (federal, provincial, and foreign) generally participated in meetings of CAFSAC. Subcommittees met in May to review groundfish stock assessments. In June, the CAFSAC Steering Committee prepared and reviewed an advisory document based on the results of subcommittee discussions. The Steering Committee consisted of an overall chairperson, the chairpersons of the seven subcommittees, the Directors of Science of the administrative regions, and a representative of the Department of Fisheries and Oceans' (DFO's) Economics Branch. The majority of the Steering Committee members had not attended the subcommittee meetings and therefore provided a second level of review. The groundfish advisory document was officially made public at a meeting of Atlantic Groundfish Advisory Committee (AGAC) in early July.

The AGAC was a joint government–industry forum to advise the Minister of Fisheries and Oceans on management measures, including total allowable catches (TACs) and quotas, for groundfish. The committee was chaired by DFO's Assistant Deputy Minister responsible for fisheries management. Provincial governments and major fishing interests were represented. The release of scientific advice after the July AGAC meeting was the starting point of consultations on scientific assessments, scientific advice, and possible management measures. After local and regional consultations were held, another AGAC meeting was held to discuss options that could be put forward to the Minister for decision and implementation. Very rarely did AGAC discussions result in consensus recommendations to the Minister. After the November AGAC meeting, DFO officials made confidential recommendations to the Minister. Interest groups had only a short time in which to influence the Minister before a decision was announced at the end of December for the fishery that would start in a few days.

The first charge of CAFSAC and AGAC was to develop an Atlantic groundfish management plan. The plan established minimum mesh sizes, TACs, quotas by stock, gear type and vessel size, time and area closures (to protect spawning and juveniles), and trip limits. The plan was relatively simple at first. It set overall TACs for each stock, quotas for mobile gears (separately for vessels longer and shorter than 34 m), and catch allowances for some gears. Catch allowances were used in fixed-gear fisheries where catch was highly variable and subject to the availability of fish, and these allowances were subtracted from the TAC before quotas were allocated. Gear sectors with allowances were not restricted even when an allowance was exceeded, but uncaught allowances were not transferred to other gear sectors. Several allowances were exceeded in the early years of the plan. This was considered unfair by gear sectors whose fisheries were closed when quotas were reached, and strong pressure from those sectors led to the elimination of all allowances by the early 1980s—except for the inshore northern cod (2J+3KL) allowance, which remained in effect until that fishery was closed in July 1992.

The Canadian Atlantic groundfish management plan grew progressively more complicated as more gear, mesh size, and seasonal categories were included each year. The process became highly confrontational, and the same individuals or the same organizations made the same arguments over the same issues year after year. By the late 1980s, the plan had reached maturity, and relatively few changes were being made from one year to the next. Everyone involved in the AGAC process realized that the results did not justify the time and energy invested in the process. They recognized that fishing pressure on several of the major stocks had to be lowered, and a 3-year plan for 1991–1993 was seen as a less frustrating and more efficient way to plan for the fishing industry.

The poor inshore fishery for northern cod during summer 1991 and the disappointing survey results the following autumn caused a reevaluation of the 3-year plan. At the end of February 1992, the Minister of Fisheries and Oceans announced a partial closure of the fishery, which became a full closure in July. At AGAC consultations during autumn 1992, scientific assessments indicated that most cod stocks in Atlantic Canada had declined sharply, that the stocks consisted of small young fish, and that very severe management measures were necessary. Recognizing that compliance with regulations had been low, that discarding had been a major problem, and that management measures had had little success in preventing discarding, the Minister required each fishing sector to prepare and obtain approval for its fishing plan for 1993. Each plan was to provide for fishery observers. When observers determined that the proportion of undersized fish exceeded 15% by number, the fishery was to be closed for a fixed period. Test fishing would take place before the fishery could be reopened. As a result, very little fishing took place in 1993.

When the Minister of Fisheries and Oceans announced (in December 1992) the management approaches for 1993, he also announced major changes in the advisory and fishery management processes. Both CAFSAC and AGAC were disbanded and a Fishery Resource Conservation Council (FRCC) was created as a partnership between government, industry, and academia to advise the Minister on conservation measures. A majority of FRCC members come from the fishing industry, but they are appointed as individuals, not as representatives of a particular sector. One DFO scientist, one DFO fisheries manager, and one representative of each of the five Atlantic provinces and the Northwest Territories are exofficio members of the FRCC. The FRCC mandate is strictly limited to advising on science matters and conservation measures. The DFO consults separately with industry on fisheries management measures.

United States

The United States undertook to manage its coastal stocks independently of ICNAF and under the aegis of the Magnuson Fishery Conservation and Management Act, which went into effect in early 1977. Under this arrangement, fishery management plans are developed by fishery management councils but must be approved by the Secretary of Commerce before they are implemented. Enforcement of fishery regulations in the extended exclusive economic zone is the responsibility of the National Marine Fisheries Service with assistance of the U.S. Coast Guard.

The New England Fishery Management Council assumed responsibility for the major share of the groundfish resources off the northeastern United States; the Mid-Atlantic Council was assigned the region south of New England. Between 1977 and 1982, management was based upon a quota system adopted from ICNAF.

Annual quotas for each stock were allocated quarterly by vessel class. Because the total allowable catches were often met or exceeded before the end of the year, quarterly allocations were further constrained by trip limits to spread the limited quota over as long a period as possible. Despite these attempts, fisheries were often closed for parts of quarters, or quota was "borrowed" from the next quarter or year.

Quota management of the New England groundfish resources was replaced in 1982 with the "Interim Plan," originally intended as a temporary measure to prevent further declines in groundfish biomass while a long-term, comprehensive management plan was being developed by the New England Fishery Management Council. The Atlantic Demersal Finfish Plan took 3 years to develop and eventually emerged in 1985 as the Northeast Multispecies Fishery Management Plan. Between 1982 and 1985 under the Interim Plan, quotas were replaced with a suite of indirect measures (primarily mesh size restrictions and seasonal closed areas).

The Northeast Multispecies Fishery Management Plan established areas in the Gulf of Maine and on Georges Bank in which trawl cod end mesh sizes had to be at least 140 mm. The minimum cod end mesh was slated to increase to 152 mm in the plan's third year, although this requirement was not implemented until 1994. Other measures codified in the multispecies plan were minimum sizes of landed fish and seasonal area closures, but a limited amount of small-mesh fishing was allowed in the Gulf of Maine.

The plan was amended seven times between 1985 and 1996. Amendments 1–4 adjusted minimum fish sizes and added overfishing definitions for most of the major groundfish stocks in the Gulf of Maine and on Georges Bank. Amendment 5, developed in response to a law suit filed by conservation groups, was implemented in 1994. It changed the original orientation of the plan by directing a 50% reduction in groundfish fishing effort over 5–7 years. In summer 1994 as well, scientists advised immediate reductions in fishing effort for several stocks. Amendment 6 (1995) expanded and extended closure areas on Georges Bank and instituted a 200-kg trip limit for haddock. Because U.S. groundfish stocks remained depressed and some, such as Georges Bank cod, continued to decline, Amendment 7 (1996) was approved to begin the process of stock rebuilding. Amendment 7 incorporates target quotas, reductions in days at sea, expansions of areas closed to fishing, changes in trip limits, and extensions of effort reduction programs to all fleet sectors, among other measures.

Scientific advice in support of management is provided to the New England and Mid-Atlantic fishery management councils through semiannual stock assessment workshops. Four assessment subcommittees and a methods subcommittee form the core of the workshops. Subcommittees assimilate assessment data provided by the National Marine Fisheries Service (NMFS), various state marine fishery agencies, and other sources; select appropriate scientific and statistical methods; and perform the stock assessments. The assessment subcommittees prepare background working papers, summary reports, and first drafts of the advisory report. A Stock Assessment Review Committee provides peer review of subcommittee assessments and prepares an advisory report to the management councils. The assessment process is guided by a Steering Committee composed of the Executive Directors of the New England and Mid-Atlantic fishery management councils, the Northeast Regional Administrator of NMFS, the Science and Research Director of NMFS's Northeast Fisheries Science Center, and the Executive Director of the Atlantic States Marine Fisheries Commission. This committee sets priorities for the assessment review process, allocates resources, and oversees the assessment advisory process. Coordination of U.S. and Canadian stock assessment and management has improved in recent years as the institutional systems evolved in both countries.

SUMMARY

Groundfish stocks off the northeastern coast of North America have undergone substantial changes in the past century, especially during the past 50 years. Prior to the arrival of large, technologically sophisticated distant-water fleets in the late 1950s, Canadian and U.S. landings of principal groundfish and flounder stocks were generally stable, and many stocks were underfished. The heavy exploitation of groundfish in the 1960s and 1970s led to drastic declines in many of the region's stocks that had heretofore supported sustainable fisheries.

After Canada and the United States extended their fisheries jurisdictions, which eliminated much foreign fishing, abundance of many stocks improved temporarily. Domestic fleets soon expanded to fill the void left by the restriction on foreign fishing, but peak landings in the early and mid-1980s were much less than during the previous 10–15 years. In the United States, groundfish landings declined markedly in the late 1980s as the long-dominant haddock and yellowtail flounder stocks collapsed. Atlantic cod on Georges Bank and in the Gulf of Maine sustained the U.S. groundfish fishery through the late 1980s and early 1990s, but both cod stocks are now in serious decline, and groundfish landings have been reduced. Off Canada, the rebound in landings was supported primarily by cod stocks off Newfoundland and Labrador, in the Gulf of St. Lawrence, and on the Scotian Shelf, as well as by development of fisheries for redfish and Greenland halibut. The decline in Canadian groundfish resources and landings was most pronounced in the 1990s, somewhat later than the U.S. experience.

The current status of many important groundfish and flounder resources of the region is poor. Many fisheries regulated by Canada and NAFO are closed outright or subject to quotas that are small fractions of historic landings (Appendix). The 12 cod stocks, which historically supplied about two-thirds of all principal groundfish landings, are in the poorest condition; all but one (Flemish Cap) exhibit low to historically low biomasses, and all but two (Flemish Cap and western Scotian Shelf) have poor to very poor recruitment prospects. Even with low fishing pressure and good recovery plans, these stocks may require 10 years or more to rebuild. Other resources are also in trouble, and the region has few "healthy" stocks.

Management bodies have responded to the resource declines by implementing more stringent regulations intended to reduce harvests of adults and to afford more protection to juveniles. In Canada, harvest quotas have been lowered to unprecedented levels–to zero in many cases. In international waters, NAFO has reduced or eliminated fishing on several important resources. In the United States, for the first time, entry of new vessels has been restricted, and controls on fishing effort have been put into place to halve fishing mortality rates. Additional restrictions intended to accelerate the pace of stock rebuilding and to cap fishing mortality rates are now being contemplated. In both countries, fisheries for alternative groundfish resources have been started or expanded. Thus, for example, landings of dogfish, skates, lumpfish, white hake, and goosefish have increased in recent years. The sustainable productivity of these resources is much smaller than the productivity of principal groundfish and flounders, and there is concern that the "new" resources may follow a similar fate, but at a more rapid pace. Programs to reduce fishing effort permanently have been undertaken in both countries in recognition that the current harvesting capacity far outstrips the long-term productivity of available resources. Permanent capacity reduction is necessary if groundfish populations are to rebuild to the point of sustaining healthy fisheries. In the United States, a few groundfish vessels were bought out through an auction conducted by the government, and additional buyouts are underway. Other vessels have left the fishery without such inducement. In Canada, fishing capacity reduction has been modest to date. Creation of a productive and sustainable groundfish fishery for the future will require a long period of rebuilding to restore depleted stocks (17).

NOTES

1. Groundfish are species that live most of their lives on or near the ocean bottom. Often called "demersal" species by fishery scientists, they can be caught with baited hooks, stationary gill nets, mobile trawls, and other bottom-tending gears.
2. Various aspects of fishing history in the northwestern Atlantic Ocean were described by Goode (1887), Alexander et al. (1915), Herrington (1935), Brown et al. (1976), Clark and Brown (1977), Clark et al. (1982), Brown and Halliday (1983), German (1987), Mayo (1987), Anthony (1990), Harris (1990), Pinhorn and Halliday (1990), Gough (1991), and Parsons (1993).
3. See Hennemuth and Rockwell (1987) and Parsons (1993) for extensive overviews of the institutional arrangements related to the management of fishery resources in the region.
4. Canadians often refer to the cod stocks off Labrador and eastern Newfoundland (subareas 2GH and 2J+3KL) as "northern cod."
5. ICNAF (1962–1972)
6. Recent Atlantic cod data for Labrador and Newfoundland stocks are from Bishop et al. (1994), Hutchings and Myers (1994b), Taggart et al. (1994), Anonymous (1995a), and (for the Flemish Cap and southern Grand Bank stocks) NAFO (1995). The history of all Canadian cod stocks was reviewed by Sinclair (1996).
7. Introduction of the otter trawl to New England groundfish fisheries and ensuing efforts to document its impacts were given by Alexander et al. (1915). An otter trawl is a non-rigid (unframed), towed net whose mouth is held open by water pressure on two "doors," one on each lateral arm of the towing bridle (see chapter 9). The switch from Atlantic cod to haddock as the primary target of groundfish fisheries in the early decades of the Twentieth Century was documented by Serchuk and Wigley (1992).

8. The buildup of the haddock fishery was rapid, and the fishery was unregulated. After the crash of haddock landings in 1929–1930, research focused on reducing high levels of discards by increasing trawl mesh sizes (Herrington 1935). Population dynamics research on the haddock stock began in the 1930s, and annual estimates of fishing mortality and stock biomass now cover seven decades for this species (Clark et al. 1982).
9. Mayo (1987) compiled a history of the U.S. redfish fishery and of research on the species' population dynamics.
10. Demand for silver hake in North American markets has historically been weak, and Canada declared harvestable surpluses of the species in its exclusive economic zone after extended jurisdiction. However, with the decline of traditional species and the increased demand in export markets such as Spain, domestic fishery interest in silver hake has increased (Anonymous 1995d, page 99).
11. Development of the American plaice fishery in the Gulf of St. Lawrence was described by Anonymous (1995a, page 49).
12. The yellowtail flounder fishery off the northeastern United States began in the 1930s in response to the decline in stocks of winter flounder, which was preferred because of its thicker fillets and more coastal distribution. The initial increase and declines in yellowtail flounder populations were described by NEFSC (1995).
13. The history of the Canadian Atlantic groundfish fishery in this century has been documented by Kirby (1982), Harris (1990), Gough (1991), Parsons (1993), Bishop et al. (1994), Chouinard and Fréchet (1994), and Anonymous (1995a, 1995b, 1995c, 1995d).
14. The U.S. history of groundfish fishing, the resource, and changes in the groundfish community as a result of fishing have been described by: Herrington (1935), Graham and Premetz (1955), German (1987), Hennemuth and Rockwell (1987), Mayo (1987), Mahon and Smith (1989), Anthony (1990, 1993), Gabriel (1992), Serchuk et al. (1994a, 1994b), and NEFSC (1995).
15. Management during the ICNAF–NAFO eras and the initial attempts at managing resources after extended jurisdiction have been recounted by Brown et al. (1976), Marchesseault et al. (1980), Kirby (1982), Brown and Halliday (1983), NEFMC (1985), Hennemuth and Rockwell (1987), NEFSC (1987, 1991, 1995), Haché (1989), Mayo et al. (1989, 1992), Anthony (1990, 1993), Harris (1990), Pinhorn and Halliday (1990), Parsons (1993), O'Boyle and Zwanenburg (1994), and NAFO (1995). The use of $F_{0.1}$ as a biological reference point for management in Canada was reviewed by Maguire and Mace (1993) and Rivard and Maguire (1993).
16. As a consequence of the World Court decision, Canada and the United States now restrict fishing by their nationals to their portions of Georges Bank. Management of transboundary stocks on the Bank is performed separately and independently by each country (Serchuk and Wigley 1992).
17. We thank Bruce Atkinson for his review of the text concerning redfishes in the western Atlantic, John Boreman and Brian Nakashima for their reviews and encouragement, and especially Bob Kendall for his very thorough and insightful editing of the original draft of the manuscript.

Appendix: Northwest Atlantic Groundfish Stocks

Table A.1.—Delineation, characteristics, and 1994 status of groundfish stocks in the northwest Atlantic Ocean (NAFO subareas 2–6). Stocks (as defined by the responsible managing authority) are named for the NAFO statistical region(s) in which they occur. Average landings are for the starting year indicated through 1993 or 1994. Data are derived from Anonymous (1995a–d), NEFSC (1995), and NAFO (1995) and historical landings records summarized in various research documents and statistical bulletins published by the two countries and NAFO. Abbreviations: N, E, S, W = north, east, south, west; NAFO = Northwest Atlantic Fisheries Organization; SSB = spawning stock biomass; YC = year-class (year of hatching).

Stocks[a]		Landings (thousands of tons)					1994 Assessment		Management responsibility	Comments
Geographic	Statistical	Average[b]	1991	1992	1993	1994	Biomass	Recruitment		
						Atlantic cod				
N Labrador	2GH	13 (1953)	0	0	0	0	Very low	Very poor	Canada	1991 survey low; possible link with 2J+3KL
S Labrador and N Grand Bank	2J+3KL	334 (1959)	171	44	11	2	Lowest observed	Very poor	NAFO	Biomass 1% of early 1980s level; YCs weak since 1987; some 1993 and 1994 YC fish
Flemish Cap	3M	26 (1960)	16	25	16	30	Moderate	Moderate	NAFO	Fishing mortality high (>1.0); 1985 and 1991 YCs highest; 1993 and 1994 YCs seem weaker
S Grand Bank	3NO	64 (1953)	29	13	10	3	Lowest observed	Poor	NAFO	Declining since mid-1980s; YCs since 1990 may be weak
N and E Gulf of St. Lawrence	3Pn+4Rs	72 (1961)	32	29	18	<1	Very low	Poor	Canada	Condition and growth poor; no signs of good recruitment
St. Pierre Bank	3Ps	47 (1959)	43	32	15	1	Low	Poor	Canada	No strong YCs after 1990; older fish gone, growth declined

Stocks[a]		Landings (thousands of tons)					1994 Assessment		Management responsibility	Comments
Geographic	Statistical	Average[b]	1991	1992	1993	1994	Biomass	Recruitment		
W Gulf of St. Lawrence	4TVn Jan–Apr	54 (1950)	49	41	5	1	Almost lowest observed	Poor	Canada	YCs poor since late 1980s; no indication of improved retention
Sydney Bight	4Vn May–Dec	8 (1970)	3	2	1	<1	Lowest observed	Very poor	Canada	Little to no recruitment since 1989; stock very depressed
Banquereau and Sable Island banks	4VsW	47 (1958)	33	30	4	<1	Lowest observed	Very poor	Canada	Recruitment poor since 1984; predation by gray seals significant
W Scotian Shelf	4X	22 (1948)	28	26	16	13	Lowest observed	Moderate	Canada	1990 and 1992 YCs about average; exploitation rate twice target
Gulf of Maine	5Y	9 (1960)	18	11	8	8	Lowest observed	Poor	USA	Spawning biomass lowest observed; exploitation rate very high
Georges Bank and south	5Z+6	33 (1960)	38	29	23	15	Lowest observed	Very poor	USA, Canada	1993 and 1994 YCs lowest observed; exploitation rate record high; Canada manages as 5Zj,m

Groundfish Stocks and the Fishing Industry 63

Stocks[a]		Landings (thousands of tons)					1994 Assessment			Management responsibility	Comments
Geographic	Statistical	Average[b]	1991	1992	1993	1994	Biomass	Recruitment			
Haddock											
Grand Bank	3LNO	13 (1953)	1	1	1	<1	Low	Poor		Canada	No signs of improved recruitment; strong 1980 and 1981 YCs fished out
St. Pierre Bank	3Ps	6 (1960)	<1	<1	<1	<1	Low	Poor		Canada	No prospects for stock improving; strong 1981 YC fished out
Central Scotian Shelf	4TVW	20 (1950)	5	6	1	<1	Almost lowest observed	Good		Canada	Stock in 4Vn and 4V's disappeared; 1992 and 1993 YCs better
Browns Bank	4X	20 (1960)	10	10	7	4	Almost lowest observed	Good		Canada	Exploitation rate above reference; adult biomass at historical low; possible 1992 and 1993 YCs improved
Gulf of Maine	5Y	3 (1956)	<1	<1	<1	<1	Lowest observed	Poor		USA	Biomass at record low; some recruits detected in surveys
Georges Bank	5Z	35 (1931)	7	6	4	3	Almost lowest observed	Poor		Canada, USA	1992 YC best since 1987; SSB increased in 1995; 1993 and 1994 YCs poor; Canada manages as 5Zj,m

Groundfish Stocks and the Fishing Industry 65

Stocks[a]		Landings (thousands of tons)					1994 Assessment			Management responsibility	Comments
Geographic	Statistical	Average[b]	1991	1992	1993	1994	Biomass	Recruitment			
Redfish											
Labrador and N Newfoundland	2+3K	25 (1960)	<1	<1	<1	<1	Very low	Poor		Canada	Virtually no recruitment since early 1970s; stock very low
Flemish Cap	3M	21 (1960)	48	43	29	11	Moderate	Moderate		NAFO	Improved recruitment expected; exploitation rate probably reduced
E Grand Bank	3LN	22 (1960)	26	27	23	7	Low	Poor		NAFO	3N contains some recruits; surveys indicate no good YCs
SW Grand Bank	3O	14 (1959)	8	17	16	4	Moderate	Improving		Canada	Large fraction of immature fish; resident versus migrant problematic
Laurentian Channel	Unit 2	24 (1960)	24	17	27	24	Moderate	Moderate		Canada	Late 1980s YC good, but not as strong as early 1980s
Gulf of St. Lawrence	Unit 1	46 (1954)	60	77	51	20	Declining	Poor		Canada	Strong 1988 YC has disappeared; survey biomass declining rapidly
S Scotian Shelf	Unit 3	7 (1970)	2	2	5	5	Moderate	Moderate		Canada	Abundance stable since late 1980s; exploitation rate believed low
Gulf of Maine	5YZ	17 (1935)	<1	1	1	<1	Low	Improving		USA	Some improvement in biomass since 1986; exploitation rate low

Stocks[a]		Landings (thousands of tons)					1994 Assessment			Management responsibility	Comments
Geographic	Statistical	Average[b]	1991	1992	1993	1994	Biomass	Recruitment			
Silver hake											
Scotian Shelf	4VWX	63 (1960)	68	32	28	8	Low to moderate	Poor to moderate		NAFO	1990 to 1993 YCs variable; 1994 weak; stock stable, lower than mid-1980s
Gulf of Maine and N Georges Bank	5YZe (part)	29 (1955)	6	5	4		Moderate	Moderate		USA	High discard of juveniles; considered fully exploited
S Georges Bank and south	5WZ(part)+6	51 (1955)	10	10	13		Low	Low		USA	Exploitation rate is very high; landings and discard of juveniles high
Pollock											
St. Pierre Bank	3Ps	1 (1967)	1	<1	<1	<1	Low	Poor		Canada	Recent surveys low; schools of small fish observed
Gulf of Maine, Georges Bank, Scotian Shelf	4VWX+5	46 (1974)	49	41	27		Low	Poor to moderate		Canada, USA	Exploitation rate high; biomass stable, lower than 1980s; Canada manages as 4VWX+5Zc
American plaice											
Labrador and NE Newfoundland	2+3K	4 (1960)	<1	<1	<1	<1	Very low	Low		Canada	Spawning stock 2% of peak; declining recruitment recently
Flemish Cap	3M	2 (1962)	2	1	<1	<1	Low	Poor		NAFO	Stock appears stable and low; 1991 and 1992 YCs appear weak

Groundfish Stocks and the Fishing Industry 67

Stocks[a]		Landings (thousands of tons)					1994 Assessment			
Geographic	Statistical	Average[b]	1991	1992	1993	1994	Biomass	Recruitment	Management responsibility	Comments
Grand Bank	3LNO	45 (1960)	34	13	17	7	Record low	Poor	NAFO	Stock declined rapidly from 1980s; no promising YCs since 1988–1989
St. Pierre Bank	3Ps	4 (1960)	4	2	<1	<1	Very low	Very poor	Canada	SSB lowest since 1975; recruitment and stock very low
S Gulf of St. Lawrence	4T	7 (1960)	8	8	2	2	Lowest since 1971	Poor	Canada	Recruitment poor late 1980s–early 1990s; no improvement
Gulf of Maine and Georges Bank	5YZ	7 (1974)	4	7	6	5	Moderate	Moderate	USA	Stock increased from mid-1980s; 1990 YC good; exploitation rate and discard high
Witch flounder										
Labrador and N Grand Bank	2J+3KL	6 (1960)	4	2	<1	<1	Very low	Poor	Canada	Distribution area shrinking; stock at extreme low level
E Grand Bank	3NO	5 (1960)	5	5	4	1	Low	Poor	NAFO	Stock near historical low; little sign of rebuilding
St. Pierre Bank	3Ps	1 (1960)	1	1	1	<1	Low	?	Canada	Recent biomass at low end of observations
Gulf of St. Lawrence	4RST	3 (1970)	1	1	1	<1	Declining	?	Canada	Large decline in abundance in 4RS; slight increase in 4T
Gulf of Maine and Georges Bank	5YZ	3 (1974)	2	2	3	3	Low	Moderate	USA, Canada	Very low biomass; strong 1990 YC; exploitation rate high

Stocks[a]		Landings (thousands of tons)					1994 Assessment			Management responsibility	Comments
Geographic	Statistical	Average[b]	1991	1992	1993	1994	Biomass	Recruitment			
Yellowtail flounder											
E Grand Bank	3LNO	16 (1965)	16	11	14	2	Low	Moderate		NAFO	Contraction in range recently; some indications of improved retention
Georges Bank	5Ze	9 (1960)	2	3	2	4	Low	Poor to moderate		USA, Canada	Exploitation rate in recent years 4 times $F_{0.1}$, 2 times overfishing definition
Cape Cod	5YZw(part)	2 (1974)	1	1	1	1	Low	Moderate		USA	Landings and surveys peaked in late 1970s; now low
S New England	5Zx(part)	10 (1960)	4	2	<1	<1	Record low	Very poor		USA	Very poor prospects for recovery; exploitation rate 4 times overfishing level
Middle Atlantic	6	1 (1974)	<1	<1	<1	<1	Very low	Poor		USA	Very poor prospects for stock rebuilding
Greenland halibut											
Labrador and Newfoundland	2+3KLMNO	23 (1960)	55–75	63	42–62	48	Declining	Good		NAFO	1990 and 1991 YCs above average; significant stock declines since 1980s
Gulf of St. Lawrence	4RST	5 (1977)	2	3	2	4	Low	Low		Canada	Low recruitment after 1988; exploitation high, targeting juveniles

Groundfish Stocks and the Fishing Industry   69

Stocks[a]		Landings (thousands of tons)					1994 Assessment			Management responsibility	Comments
Geographic	Statistical	Average[b]	1991	1992	1993	1994	Biomass	Recruitment			
Atlantic halibut											
Grand Bank and Scotian Shelf	3NOPs+4VWX	2 (1970)	2	1	2	1	Stable	Stable		Canada	Total effort stable since 1992
Gulf of St. Lawrence	4RST	<1 (1980)	<1	<1	<1	<1	Stable	Good		Canada	Signs of good recruitment; 60% of catch smaller than 81 cm (maturity)
Winter flounder											
Gulf of St. Lawrence	4T	2 (1970)	3	2	1	1	Average	?		Canada	Landings level uncertain; overall abundance about average
Gulf of Maine	5Y	4 (1974)	1	1	1		Low	Poor		USA	Historical low landings; survey indices near lowest
Georges Bank	5Ze	3 (1974)	2	2	2		Low	Poor		USA	Continuous decline since late 1970s; overexploited
S New England and south	5Zw+6	9 (1974)	6	4	4		Low	Poor		USA	Landings and survey indices at historical lows

Stocks[a]		Landings (thousands of tons)					1994 Assessment		Management responsibility	Comments
Geographic	Statistical	Average[b]	1991	1992	1993	1994	Biomass	Recruitment		
Flatfish complex										
Central Scotian Shelf 4VW (plaice, witch, winter, yellowtail)		12 (1970)	5	5	4	3	Very low	Increasing for some	Canada	Plaice and witch low; signs of recruitment; yellowtail very low
Browns Bank (plaice, 4X witch, winter, yellowtail)		4 (1970)	6	6	4	3	Moderate	Moderate for some	Canada	Plaice stable; yellowtail increasing; witch low with signs of retention; signs of recruitment for plaice

[a] See Figure 2.1 for NAFO statistical areas.
[b] The year in parentheses is the first year of consistent records for the stock. Average landings were calculated from that year through 1994, except 1993 is the concluding year for silver hake stocks 5YZe and 5Zw+6, pollock 4VWX+5, American plaice 5YZ, witch flounder 5YZ, yellowtail flounder (all stocks), and winter flounder 5Y, 5Ze, and 5Zw+6.

Alan F. Sinclair and Steven A. Murawski

WHY HAVE GROUNDFISH STOCKS DECLINED?

Groundfish fisheries in the northwest Atlantic are in an unprecedented state of decline, as described in Chapter 2. The current problem involves almost all of the commercially important groundfish species of the region, including the dominant Atlantic cod plus haddock; pollock; American plaice; yellowtail, witch, and winter flounders; redfish; and others. These species have been the object of extensive management and research for decades, and considerable long-term information about them is available.

In this chapter, we review the available information on factors that might have influenced the declines in these resources and their fisheries (1). We have grouped these factors into four categories: (a) biological considerations, (b) environmental considerations, (c) species interactions, and (d) fishing.

OVERVIEW OF GROUNDFISH ABUNDANCE

Groundfish landings are recorded, and fish stocks typically are named and managed, according to a geographic system established by the International Commission for the Northwest Atlantic Fisheries (ICNAF) in the late 1940s. This system was retained by ICNAF's successor, the Northwest Atlantic Fisheries Organization (NAFO), and it remains in effect today (Figure 3.1). The NAFO "convention area" extends from near the Arctic Circle south to the latitude of Cape Hatteras, North Carolina (35°N), and east to longitude 42°W. Within this area are seven subareas, numbered from north to south. Subarea 0 covers waters off Ellesmere and Baffin islands in Canada; subarea 1 is west and south of Greenland; subarea 2 is east of Labrador; subarea 3 is east and south of Newfoundland (including the Grand Bank and Flemish Cap); subarea 4 includes the Gulf of St. Lawrence, Bay of Fundy, and waters east and south of Nova Scotia; subarea 5 is east and south of

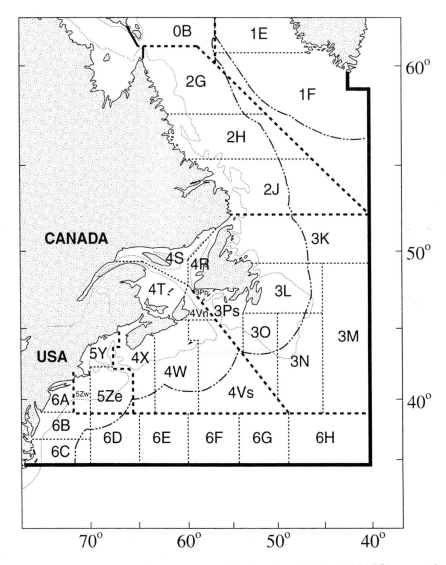

FIGURE 3.1.—Map of the Northwest Atlantic Fisheries Organization's (NAFO) convention area (heavy solid line). Subareas are numbered 0 to 6 from north to south (heavy broken lines). Divisions within subareas have capital letters that proceed alphabetically from north to south (light broken lines). Subdivisions have lowercase letters denoting compass direction (n, e, s, or w). The light dotted line indicates the 183-m (100-fathom) depth contour. The dotted–dashed line marks the limit of national jurisdictions (370 km or 200 nautical miles from shore). (Map courtesy of K. Drinkwater and D. Mountain, from Chapter 1.)

New England (including Georges Bank); and subarea 6 embraces waters off the mid-Atlantic states and along the area's southern tier. Subareas are further broken down into divisions and subdivisions. The large majority of groundfish landings have come from subareas 2–5 (Labrador to New England).

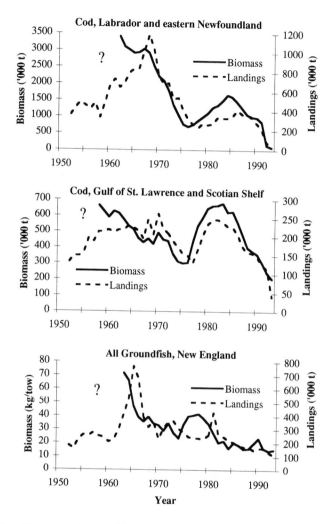

FIGURE 3.2.—Correspondence of landings and biomass estimates for Atlantic cod or all groundfish in three areas of the northwest Atlantic, 1952–1993. Data are in thousands of metric tons, except New England biomass estimates are in kilograms per trawl tow. Labrador and eastern Newfoundland data are from NAFO subareas 2 and 3, Gulf of St. Lawrence and Scotian Shelf values are from subarea 4, and New England data are from subareas 5 and 6. Question marks indicate periods when biomass was not estimated.

Indexes of groundfish abundance (typically estimated as aggregate weight, or biomass) have been derived annually by Canadian and U.S. biologists since the early 1960s. During this period, biomass indexes have correlated quite well with commercial groundfish landings; examples are shown in Figure 3.2. Landings may be imprecise indicators of abundance over short time periods (5 years) because changes in catch are related to fishing effort. A substantial increase in landings could mean an increase in abundance or an increase in fishing pressure on lightly

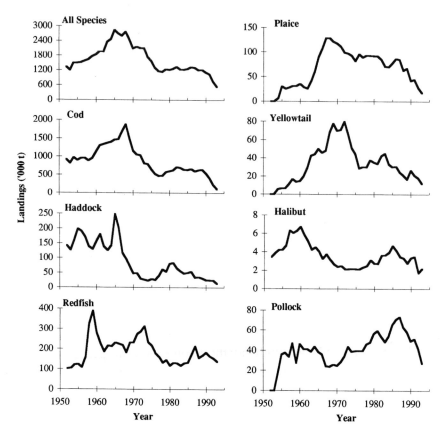

FIGURE 3.3.—Groundfish landings (thousands of tons) throughout the NAFO area, 1952–1993. Plots are shown for an aggregate of all species and separately for Atlantic cod, haddock, redfish (two or three species), American plaice, yellowtail flounder, halibut, and pollock.

fished stocks. However, a long-term downward trend in landings is likely to indicate decreasing fish abundance, especially if fishing effort has remained relatively constant or has increased.

The combined landings of all groundfish species within the NAFO convention area were relatively stable in the mid-1950s, varying between 1.3 million and 1.5 million tons annually (Figure 3.3). Landings then almost doubled to about 2.7 million tons in 1965. This was a period of expanding distant-water (non-North American) fishing effort, which continued until the mid-1970s (2). After relatively few years at this high level, landings declined rapidly to 1.1 million tons in 1978. The groundfish fleet underwent a transition from largely distant-water vessels to predominantly domestic ones with the 1977 implementation of extended fisheries jurisdictions in both the United States and Canada. Annual landings stabilized during the 1980s at around 1.3 million tons, slightly below those of the mid-1950s, but then declined sharply beginning in 1990. The large Atlantic cod fishery extending from southern Labrador to southeast Newfoundland (divisions 2J, 3K, and 3L) was closed in July

1992, and most other cod and many other groundfish fisheries in Canada were closed in 1993. The only cod fisheries that continued to operate in 1993 were off Nova Scotia (division 4X) and New England (subarea 5). In New England, certain areas (but not entire fisheries) were closed in 1994 and 1995.

Cod has dominated the groundfish landings, accounting for 54% during 1952–1993. During the 1950s, cod landings were relatively stable, close to 1.0 million tons annually (Figure 3.3). Landings began to increase in the late 1950s and continued to a peak of 1.9 million tons in 1968. Subsequently, there was a rapid decline to a minimum of 0.5 million tons in 1977. Cod landings increased slightly in the years following the extension of fisheries jurisdiction and varied between 0.6 and 0.7 million tons annually during the 1980s, substantially below the volume landed in the 1950s. Landings declined again in the late 1980s and reached the lowest levels on record in the early 1990s.

Haddock landings varied between 130,000 and 200,000 tons annually during the 1950s and early 1960s (Figure 3.3). A peak of haddock landings occurred in 1965 at 250,000 tons, which was followed by a steady decline to 23,000 tons in 1974. Landings subsequently increased to 81,000 tons in 1981 and have declined steadily since then to a minimum of 13,000 tons in 1993.

Annual landings of redfish have varied between 100,000 and 400,000 tons since the early 1950s; three peak periods corresponded to episodes of good recruitment (Figure 3.3). The first peak in the late 1950s was dominated by landings off Labrador and eastern Newfoundland; the second and third peaks were driven by landings in the Gulf of St. Lawrence, Scotian Shelf, and Gulf of Maine. The dominant trend indicated by sequential peaks is downward.

American plaice was the most abundant flatfish species landed in the NAFO area until recently, when it was replaced by Greenland halibut (turbot). The plaice fishery developed in the mid-1960s, when total landings increased from about 25,000 to 130,000 tons (Figure 3.3). Total landings have declined steadily since the late 1960s to a level as low as those of the 1950s. The fishery for yellowtail flounder also developed during the 1960s (Figure 3.3). Landings increased steadily between 1954 and 1972 to a peak of 80,000 tons. This was followed by a general downward trend to a level comparable to those in the 1950s.

Landings of Atlantic halibut have been small, but the species is highly valued. Total landings peaked in 1960 at 6,700 tons, then declined to about 3,000 tons in the mid-1970s (Figure 3.3). They increased somewhat until 1985, but then declined again to the lowest on record in 1992–1993.

The pollock fishery has been concentrated in the southern part of the NAFO area. Annual total landings varied between 20,000 and 40,000 tons during the 1950s and 1960s (Figure 3.3). Then they increased to a peak of 73,000 tons in 1987, but fell to 27,000 tons in 1993.

In summary, groundfish abundance in most of the NAFO area has shown a long-term downward trend from the 1950s to the present. Recent declines in landings and abundance have been precipitous, reaching the lowest levels recorded. Some species declined more rapidly and to a greater degree than others, leading to changes in the species composition of landings. Although cod, haddock, and redfish were commercially important in the 1950s, fisheries for flatfish (American plaice and yellowtail flounder) increased in the 1960s and fisheries for pollock increased throughout the 1970s and 1980s. To examine the factors responsible for these trends, we have considered two time scales, a long term spanning four decades and a shorter period from the mid-1980s to the present. In the latter case, we have focused on factors that could have affected the survival of year-classes spawned in the mid- to late 1980s, year-classes that should have supported the fisheries in the early 1990s.

BIOLOGICAL CONSIDERATIONS

Key biological characteristics of the major groundfish stocks in the area have changed. Possible causes of these changes and their potential influences on the decline of groundfish resources are discussed below.

Spawning Components

Important spawning components have virtually disappeared from two cod stocks in the NAFO area. Off Nova Scotia, spring spawning by the 4VsW stock (the stock spanning subdivision 4Vs and division 4W; Figure 3.1) has been lost (3). Surveys of eggs and larvae in the late 1970s and early 1980s had revealed both spring and late-fall spawning, but only fall reproduction was detected in the early 1990s. This observation was supported by a reduction in length of age-1 cod in the area, indicative of a dominance of fall-spawned fish (4), and commercial fishers abandoned the traditional spring spawning grounds and moved to an over-wintering area. Offshore spawning concentrations of cod nearly disappeared off southern Labrador (division 2J) in the 1990s. Commercial concentrations of cod were found only off eastern Newfoundland (divisions 3KL) during the 1990s, whereas large concentrations had previously existed in division 2J. Waters in this region have become colder. Whether or not the population shifted its distribution or simply disappeared from 2J and whether this change was caused by lower water temperatures remain controversial (5). Regardless of the cause, reduced spawning activity is likely to result in reduced stock abundance.

Weight at Age and Condition

Weight at age is a measure of growth; it is the average weight of fish of a given age (0, 1, 2, etc.). The greater the weight at age, the faster fish have grown up to that age. All other things remaining equal, faster-growing fish produce more yield to a commercial fishery and have a higher reproductive capacity; bigger fish produce more eggs. Thus, weight at age is an important statistic for assessing stock dynamics.

Weights at age of several cod stocks from southern Labrador to eastern Nova Scotia declined from relatively high values in the late 1970s to relatively low values in the mid-1980s (6). Low weights at age persisted in the late 1980s and early 1990s. The 1992 weights at age 7 for these stocks were between 52 and 70% of those in 1978. This contributed to a reduction in the biomass (total weight) of spawning fish; for example, the spawning biomass of the southern Gulf of St.

Lawrence (division 4T) cod in 1992 would have been 1.8 times higher if the population had the same weights at age as in 1978. No trends were evident for the more seaward and southern stocks (3NO, 3Ps, 4X, and 5Z).

Several possible causes can be cited for these variations in weight at age. The highest weights at age occurred in a period of low cod abundance (late 1970s), and then declined when the populations increased during the early 1980s. This suggests that growth rates may be related to population size, being highest when abundance is low, and vice versa (7). However, low weights at age have persisted during the recent period of low abundance. Fish grow more slowly when water temperatures are low and some biologists have reported relationships between growth and temperature in the wild (8). Recent temperatures in the northern areas have been below average, but the decline in weights at age occurred when temperatures were above average. A drawback of these studies is that temperature measurements are most often available for broad geographic areas and little work has been done with temperatures of water actually occupied by the fish. Some of the decline in size at age of cod in the southern Gulf of St. Lawrence (division 4T) has been attributed to selective fishing of the largest fish of a particular age-class (9). However, the declining trends did not intensify when fishing increased in the late 1980s, nor have weights at age significantly improved since the closure of the fishery in 1993. At this time there is no clear explanation for these trends in weight at age, only several possible hypotheses. The true cause may be a combination of two or more influences.

Fish condition is a measure of "well being." An index commonly used is the ratio of total weight to the length cubed. Another relative index is the weight of a fish of a given length predicted from a statistical relationship between the two variables (referred to here as weight at length). Other indexes of condition focus on specific organs such as the liver or the gonads, and these may be calculated as the ratio of organ weight to the weight of gutted fish. Good condition means fish are feeding well and have built up energy reserves to get them through winter, reproduction, migrations, and other energetically demanding periods. Poor condition may indicate that the fish are starving and under stress.

There is little to suggest that cod condition has changed in a manner that could explain their recent decline in abundance. In the Gulf of St. Lawrence, the highest condition of Atlantic cod is attained in the fall following a summer of intensive feeding. Minimum levels occur in late spring just after spawning. In the southern Gulf of St. Lawrence (division 4T), fall weight at length has had no consistent trend between 1971 and 1994, whereas spring sampling conducted in 1991–1994 indicated a minimum in condition in 1992 and an increase since (10). Condition of cod in the northern Gulf of St. Lawrence (divisions 4R and 4S) during May 1994 was near levels that were lethal in laboratory experiments (11); perhaps higher-than-normal starvation mortality occurred, but this was well after the stock had collapsed. Condition appeared to recover in 1995. Recent declines (during the 1990s) in mean weight of 35- and 60-cm cod have been reported off eastern Nova Scotia (division 4VsW). To the north, fall cod weight at length off Labrador (division 2J) and eastern Newfoundland (division 3K) declined from 1989 to 1992 but then recovered; no trend was observed in the adjacent division (3L). However, these

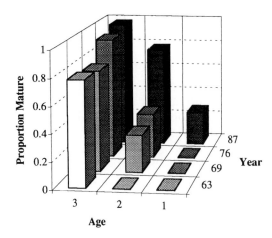

FIGURE 3.4.—Changes in age at maturity of haddock on Georges Bank. In 1963, the youngest fish to mature were age 3. In 1969, more than one-fourth of age-2 fish matured. By 1987, even some age-1 fish were reproductively mature.

latter trends may be within the range of normal variation (12). In any event, the reported changes in condition occurred in the 1990s, after the decline in cod abundance was well underway.

Cod feeding has been measured during most Canadian fall surveys off Labrador (division 2J) and eastern Newfoundland (division 3K and 3L) since 1989 (13). Stomach fullness decreased in division 2J during 1991–1994 and in 3K in 1994, mainly because capelin almost disappeared from the cod diet. These declines in feeding, however, came after, not before, the declines in cod abundance. Moreover, the feeding data do not correspond with trends in condition.

Changes in Maturity

Several groundfish species in waters off New England (subareas 5 and 6) have matured at progressively smaller sizes and younger ages since the late 1960s; an example is shown in Figure 3.4 (14). Growth rates of haddock in this area have also increased during the same time period. The age at maturity of cod off Labrador and eastern Newfoundland (divisions 2J, 3K, and 3L) was relatively stable throughout the 1980s, but declined in the early 1990s (15). Earlier maturity and faster growth might reflect higher per-capita prey rations as predator abundance decreased. Alternatively, early maturing fish may have been favored by intense fishing pressure on older adults. In any case, changes in age at maturity are most likely a consequence, not a cause, of declining groundfish abundance.

ENVIRONMENTAL CONSIDERATIONS

Environmental changes are likely to affect fish populations in three main ways: by improving or reducing the survival of young, by influencing fish growth and therefore productivity, and by altering the spatial distribution of adults. Freshwater runoff from land, ocean salinity, water temperature, and atmospheric pressure have been the environmental variables most often examined for their possible effects on groundfish populations (16).

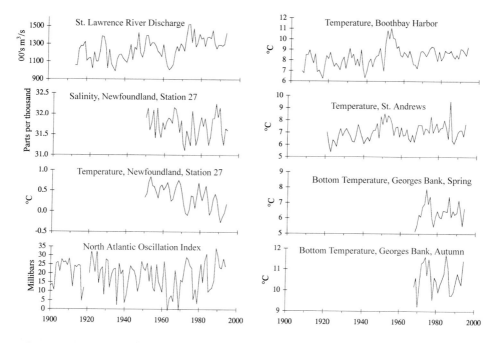

FIGURE 3.5.—Twentieth Century trends in St. Lawrence River discharge (hundreds of cubic meters per second), salinity and temperature off Newfoundland (station 27 is east of St. John's), the atmospheric pressure difference between Iceland and the Azores (North Atlantic Oscillation index), surface temperatures in the Gulf of Maine (Boothbay Harbor) and Bay of Fundy (St. Andrews), and bottom temperatures on Georges Bank in spring and autumn.

Nearly all studies have been based on correlation analysis, in which changes in an environmental variable over a sequence of seasons or years are matched with changes in the population over a comparable time period (17). A correlation that is consistently "positive" (both population and the environmental variable increase together) or "negative" (the population decreases when the environmental variable increases, or vice versa) over time suggests that the environmental factor affects the population in some way. "Suggests" is an important qualifier, because additional evidence is required to prove that a correlation is not just a coincidence. It often happens that a correlation that looks strong for a few years breaks down as more years go by.

Runoff from land brings nutrients to the sea, and changes in nutrient availability affect the food chains leading to fish and other harvested species. Discharge of the St. Lawrence River, the single most important river in the NAFO area, decreased during the 1950s, reached a minimum in 1965, then increased to a maximum in the mid-1970s. Since then, it has fluctuated without apparent trend (Figure 3.5). Analyses published during the 1970s showed good correlations between St. Lawrence River discharge and harvests some years later of American lobster, soft-shell clams, and halibut in the southern Gulf of St. Lawrence and of 10 commercial species in the Gulf of Maine (18). (Water from the St. Lawrence River makes its way through both regions; see Chapter 1.) However, these correlations had mixed success in predicting landings over the next decade; they worked for American lobster and soft-shell clam—but not for halibut—in the Gulf of St. Lawrence, and they failed for several species (including Atlantic cod) in the Gulf of Maine (19). We did a new correlation analysis to relate St. Lawrence River discharge over the years 1954–1989 to cod landings throughout the NAFO area 4 years later. The overall correlation was quite strong (and negative), but it failed to account for the dramatic decline in cod landings since 1990. Moreover, regional correlations were strongest for landings in northern subareas 1–3, on which the St. Lawrence River has little or no influence, and weak for subareas 4 and 5, which are downstream from the river mouth. Thus, the influence of freshwater runoff on groundfish harvests is ambiguous at best.

Changes in salinity are often associated with changes in nutrient concentration and thus in biological productivity because they reflect changes in freshwater runoff, the formation and melting of sea ice, and the intrusion and mixing of oceanic water masses. Salinities have been measured since 1954 at sampling station 27 off St. John's, Newfoundland (20), where they have varied over a range of 2 parts per thousand without long-term trend (Figure 3.5). Three attempts have been made to correlate summer salinity at station 27 with the number (0–4 years later) of age-3 and age-4 Atlantic cod around Newfoundland. The abundances of young fish were estimated from commercial landings. The first attempt, covering the years 1958–1976, produced a very strong positive correlation between salinity and fish abundance (21). The second, which extended the analysis to 1987, yielded a good but weaker correlation (22). The third, a reanalysis of the second after errors in abundance estimates had been corrected, produced no correlation with salinity (23).

Fish species are adapted to particular ranges of temperature, and marked changes in water temperature can force changes in their geographic distributions. In the southern part of the NAFO area, water temperatures have varied over several degrees centigrade without strong long-term trends. Surface waters in the Gulf of Maine (measured near Boothbay Harbor, Maine) and the lower Bay of Fundy (St. Andrews, New Brunswick) warmed from 1940 into the latter 1950s, then cooled into the 1960s; after a slight increase, temperatures have been relatively consistent since 1970 (Figure 3.5). Middepth temperatures in the Gulf of Maine and over the Scotian Shelf, however, have cooled considerably since the mid-1980s. Bottom temperatures of U.S. waters on Georges Bank were cold in the 1960s but have varied without trend since. Year-to-year variation in temperatures of this region may have a pronounced influence on the distribution of some stocks, but few of them are groundfish (24). The first effort to relate harvests to sea surface temperature for the Gulf of Maine was encouraging; landings of warmwater fish and shellfish were positively correlated with temperature (landings and temperature increased and decreased together) and catches of coldwater species were negatively correlated (landings decreased when temperature increased and vice versa), which was consistent with expectations (25). These relationships did not accurately predict subsequent harvests of Atlantic cod and several other species, however (26). Farther north at station 27 off Newfoundland, the temperature trend has been downward since 1950, reaching a minimum in 1991 (Figure 3.5). These waters, under the strong influence of the Labrador current, are always cold, and the long-term decline has been less than 1.5°C. As with salinity, temperature at station 27 has shown no correlation with the abundance of juvenile cod (27).

Thus, efforts to define environmental controls on groundfish populations have generated initial enthusiasm followed by gradual disillusionment. Yet, there are strong hints that environmental forces have influenced recruitment—the number of young fish that enter the fishable populations each year—even if those forces cannot presently be identified. Recruitment to Atlantic cod and haddock fisheries seems to have varied synchronously over broad geographic regions regardless of fishing pressure or the sizes of the stocks themselves. Scientists once thought that recruitment of both species varied together over the entire NAFO area. Then they realized first, that cod and haddock recruitment varied independently of one another and second, that recruitment patterns of both species differed north and south of mid-Nova Scotian waters (roughly the boundary between divisions 4W and 4X; Figure 3.1) (28). These findings are reasonable in several respects. Fish, like any other wild animals, do react and adjust to their environments; if environments vary over broad regions, so should fish stocks. Different species (such as cod and haddock) often react differently to the same environmental changes. Oceanic transitions do occur off Atlantic Canada, where the influences of the Gulf Stream and southern atmospheric conditions finally disappear and those of the Labrador Current and subboreal climates become dominant (Chapter 1). Still, the causes of these recruitment synchronies are not presently understood. The North Atlantic Oscillation (NAO) index (Figure 3.5), the difference in atmospheric pressure between the typical highs near the Azores Islands and the lows near Iceland, integrates climatic processes that have occurred over the northwest Atlantic (Chapter 1). A clue to the dynamics of northwestern Atlantic groundfish stocks may be hidden in the NAO index, but it has not yet been found (29).

Major efforts to document the responses of groundfish to changes in their environments have brought ambiguous results. Of the many environmental variables described here and in Chapter 1, only temperature at station 27 shows a long-term (downward) trend over the same period that groundfish landings have declined (1950s to the present). However, their variations on shorter time scales do not match. If the environment was mainly responsible for the recent crash in groundfish stocks, one would expect recent values of environmental variables to be extreme and present for a number of years. Such a pattern is not evident in any of the environmental variables examined. We conclude that the influences of environmental variables on Atlantic groundfish stocks have been overwhelmed by other forces, and that they cannot, by themselves, account for the general declines in groundfish harvests over the past 40 years or the precipitous declines that have occurred since 1990.

SPECIES INTERACTIONS

Competition among species may lead to important changes in ecosystems. Introductions of new species to relatively isolated ecosystems have resulted in well-known extinctions or massive reductions of native species. The effects may be direct, such as through increased predation, or indirect, such as by reductions in important prey species or competition for space. One question of interest to us is whether competition among species caused the declines in abundance of traditionally fished species of northwest Atlantic groundfish. Another important question is whether the gaps left by these declines have been filled by other species, thus impeding recovery of the depleted stocks. One way to address these questions is by analyzing changes in species composition on the fishing grounds.

Species Composition

Fish assemblages in the northwest Atlantic have changed to various degrees over the past 25 years, and the degree of change has increased from north to south. Groundfish groups with relatively homogeneous and persistent species composition have been identified on the Grand Bank (30). During 1971–1987, Atlantic cod became more numerous and thorny skate less so. American plaice abundance decreased in the northeastern region, but not elsewhere; overall, species composition was generally stable. More recently, the abundances of lightly exploited species such as American plaice, witch flounder, and capelin, as well as other noncommercial species, have declined off Labrador and eastern Newfoundland (divisions 2J and 3K). Reasons for these declines are unclear. Large-scale but still unquantified environmental changes might have been influential, but commercial bycatches of the groundfish among these species may also be important (31). Conversely, Arctic cod, crabs, and shrimp seem to be increasing in northern waters. On the Grand Bank (divisions 3L, 3NO, and 3O), however, there still is little sign that other species have increased in abundance to fill the gaps left by declining groundfish species.

An analysis of the groundfish survey catches on the Scotian Shelf (divisions 4V, 4W, and 4X) during 1970–1981 indicated very little change in species composition and distribution. Nine different groups of species in 10 areas were defined, and they persisted throughout the period of study with little variation (32). Capelin has increased in abundance on the eastern Scotian Shelf during the 1990s as bottom temperatures decreased (33). Capelin are normally found in more northern and colder waters.

Major changes in bottom fish assemblages have occurred in the Gulf of Maine, Georges Bank, and the Middle Atlantic Bight. In the Gulf of Maine, species diversity remained relatively stable, but deepwater species such as redfish and witch flounder declined dramatically from 1963 to 1988, whereas goosefish and white hake increased (34). On Georges Bank, species diversity decreased steadily throughout the 1970s and 1980s; commercial species such as haddock and yellowtail flounder declined most precipitously. In later years, winter skate, longhorn sculpin, and spiny dogfish increased dramatically, particularly as a percentage of the overall Georges Bank bottom fish assemblage. In the northern Middle Atlantic Bight, diversity was variable but increased throughout the period, whereas diversity in the southern bight was stable. Increases in the abundance of dogfish, squid, butterfish, fourspot flounder and searobins were the most apparent.

The dramatic decline in U.S. groundfish abundance (subareas 5 and 6) has been accompanied by changes in other fish components (35) (Figure 3.6). In particular, small elasmobranchs (spiny dogfish and skates) have doubled in abun-

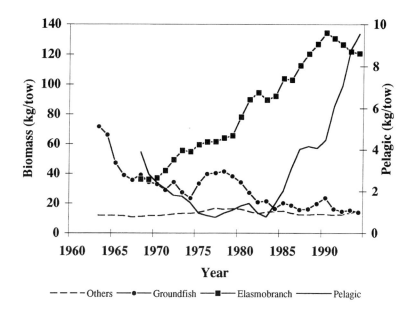

FIGURE 3.6.—Changes in biomass of four species groups in U.S. waters (subareas 5 and 6), 1963–1994. The data are from spring and autumn bottom trawl surveys. "Groundfish" include Atlantic cod, haddock, and the principal flounders caught in autumn surveys. "Pelagic" species include Atlantic herring and Atlantic mackerel taken in spring surveys. "Elasmobranchs" are spiny dogfish (a shark) and skates caught in autumn. "Others" includes goosefish, scup, cusk, weakfish, and miscellaneous species taken in autumn. The biomass scale for pelagic species (right side of graph) differs from that of the three other groups. Most pelagic fish swim too high in the water column to be captured by the bottom trawls.

dance since 1970. Spiny dogfish have accounted for most of the increase in elasmobranch abundance, particularly since 1980. Skate abundance, although near record high levels (for the aggregate of seven species), has remained relatively stable since the late 1980s (36). Atlantic herring and Atlantic mackerel have quadrupled in number since 1985. Atlantic herring were fished to extinction on Georges Bank in the mid-1970s and it took 15 years of no fishing before the stock showed signs of recovery. Mackerel have also been lightly exploited over the same period.

Increases in pelagic and elasmobranch stocks as groundfish declined raise the question of interactions among the species. Feeding studies indicate that most of the fish eaten by spiny dogfish are pelagic species; in the summers of 1984–1987, less than 7% (by volume) of the prey consumed by spiny dogfish on Georges Bank were cod, haddock, or flounders (37). We conclude that dogfish predation on groundfish juveniles is probably not an important source of mortality. Predation by herring and mackerel on eggs and larvae of groundfish is possible, but there is no evidence that it could explain the decline in groundfish resources in subareas 5 and 6.

Will these abundant species impede the recovery of groundfish in subareas 5 and 6? Competition by skates and dogfish with groundfish is apparently weak, because growth rates of individual groundfish are currently at or near record high levels for many stocks on Georges Bank. Groundfish and elasmobranch diets are not very similar, and most bottom species are opportunistic feeders, eating whatever prey is most abundant. As already noted, dogfish seem to feed only incidentally on juvenile groundfish, and even this predation is seasonal owing to the extensive migrations of dogfish. Moreover, cod and haddock rapidly grow out of the size range consumed by this predator. Thus, it is unlikely that initial recoveries of groundfish resources in subareas 5 and 6 will be impeded by the elasmobranch stocks, although the degree to which the ecosystem can support high biomasses of both elasmobranchs and other groundfish remains conjectural. There is no evidence for significant predation by herring and mackerel on groundfish eggs and larvae, although studies are currently underway to evaluate potential interactions.

North of subareas 5 and 6, there has been little change in community structure despite large declines in groundfish abundance.

Marine Mammal Predation

Considerable attention has been paid recently to the potential impact of seal predation on Atlantic cod populations in Canada. To date, most of the work has focused on grey and harp seals (38).

Breeding populations of grey seals occur on Sable Island on the Scotian Shelf (near the boundary of divisions 4V and 4W) and in the southern Gulf of St. Lawrence. The Sable Island population was estimated to have 81,600 individuals in 1993 and to be growing at a rate of 13% a year, a doubling time of 6 years. The Gulf of St. Lawrence population was estimated to have 61,900 individuals in 1993 and a growth rate of 10% per year. An analysis of stomach contents indicated that 13% of the grey seal diet was Atlantic cod. The total consumption of Atlantic cod by the two grey seal populations was estimated to be in the range of 38,000–43,000 tons in 1993. Most of the cod eaten were smaller than 40 cm long (the minimum acceptable commercial size) and were 2 and 3 years old.

Harp seals migrate between summer feeding areas in the Arctic and two winter breeding areas, one in the Gulf of St. Lawrence and the other off northern Newfoundland. The total number of harp seals was estimated to be 4.8 million in 1994; it has been increasing at 5% per year since the early 1970s. Prey consumption by harp seals in 1994 was estimated to be 6.9 million tons, 46% of which was taken in the Arctic, 40% off eastern Newfoundland, and 14% from the Gulf of St. Lawrence. Off Newfoundland, Arctic cod and capelin accounted for 65% of the diet. Atlantic cod consumption, 3% of the diet, was estimated to be approximately 88,000 tons, almost twice the amount (46,000 tons) consumed in 1981. In the Gulf of St. Lawrence, about 48% of the harp seal diet was capelin and Arctic cod. Atlantic cod made up 5.6% of the diet and the 1993 consumption was estimated to be 54,000 tons, again a near doubling since 1981 (28,000 tons). Most of the cod consumed were smaller than 40 cm and the majority were 10–20 cm long.

It is not possible to estimate total pre-recruit mortality of Atlantic cod due to seal predation, but seal populations have been increasing over the past two decades, so their total consumption of cod has probably also increased. Current estimates of cod recruitment at ages 1–3 are less than half those from the early 1980s, suggesting that the impact of seal predation may be growing. If seal predation accounts for a substantial source of juvenile cod mortality, then recovery of the cod resources may be delayed even in the absence of fishing. However, it is unlikely that seal predation caused the recent decline in cod stocks. Adult cod abundance declined before that of pre-recruits in most northern areas. One would expect the opposite pattern if seal predation were an important factor, because seals prey on pre-recruit age-groups.

Fish consumption by whales and dolphins in subareas 5 and 6 may exceed 120,000 tons annually, but most of it is pelagic fish plus some silver hake (39). These marine mammal stocks will have little direct influence on groundfish recovery rates, and they are not implicated in groundfish declines.

FISHING

Fisheries management measures are established to sustain productive fisheries—productive in terms of both harvests and biological reproduction. Managers try to prevent two kinds of excessive fishing. One, called "growth overfishing," arises when fish are caught at too young an age and the benefits of their future growth are lost. More individuals are caught, but their average weight is low and the total yield (harvested weight) from the fishery is less than it could have been had the young fish been allowed to grow. The other kind, "recruitment overfishing," occurs when the spawning stock is exploited so heavily that reproduction is depressed and subsequent recruitment of young fish to the population is too low to support a viable fishery.

Managers attempt to prevent overfishing by controlling the amount of mortality caused by fishing. The most common index of mortality used in fisheries management is the "instantaneous rate of fishing mortality" (F), which is commonly referred to simply as "fishing mortality" (40). Two mortality "targets" have been used to prevent growth overfishing. One—F_{MAX}—is the fishing mortality that produces the maximum yield from each unit of new recruits. The other—$F_{0.1}$—is the mortality at which an additional small increase in fishing effort will bring only 10% of the yield per unit of recruitment that the same increase in effort would bring from an unfished population (41). Fishing at F_{MAX}, by definition, maximizes yield per recruit, but attaining the last increments can be expensive. Normally, the yield at F_{MAX} is about 10% higher than the yield at $F_{0.1}$; however it may take 60% more fishing effort to get it. Fishing at $F_{0.1}$ is more profitable than fishing at F_{MAX}.

Yield-per-recruit models, which are used to estimate these target fishing mortalities, were adopted for northwest Atlantic groundfish during the ICNAF period, and F_{MAX} was adopted as the target. These models were used primarily to prevent growth overfishing, but the target fishing mortalities were judged low enough to prevent recruitment overfishing as well, although relationships between recruitment and stock size were too poorly understood at the outset to confirm this. Management was implemented with catch quotas set at levels estimated to generate the target fishing mortality (42).

Yield-per-recruit modeling was continued by both Canada and the United States following the extensions of fisheries jurisdiction in 1977. In Canada, $F_{0.1}$ was adopted as the management target for most resources to better protect the spawning potential of the stocks and to make the fisheries more economically efficient (43). No explicit minimum biomass thresholds have been defined, but fisheries have been restricted or, more recently, closed when stock size was critically low.

86   Why Have Groundfish Stocks Declined?

The United States retained F_{MAX} as the basis for quota management through 1981 but gave special protection to severely depleted stocks such as Georges Bank haddock. Quota-based management of groundfish proved politically difficult in the United States and it was abandoned in 1982 (44). When asked about the potential for overfishing the resource in the absence of direct controls, one manager suggested that if a resource crash occurred "it would at least prove clearly that it is possible to overfish and that there are valid reasons for resource management. On the other hand, if fishers continue to do very well, the assessments would be proven wrong once and for all" (45).

Since 1982, U.S. scientists have evaluated ways to determine optimal harvest rates for spawning biomass replacement (46). The basic idea is that if a year-class is not fished at all, it will produce the maximum amount of spawning biomass and it will realize its maximum spawning potential. As fishing mortality increases, fewer fish of the year-class spawn multiple times because they are caught in the fishery, and the spawning potential is reduced. If a stock is to sustain itself, year-classes must produce at least as much spawning biomass as existed in their year of birth. Target fishing mortalities are chosen so that harvests will not exceed a percentage—generally in the range of 20–35% of the maximum spawning potential. Resource conservation is an explicit goal of these models, and they now form the basis for defining recruitment overfishing in the United States.

Trends in Fishing Mortality

Most northwest Atlantic groundfish stocks endured rising fishing mortality during the 1960s and early 1970s, primarily the result of intensive fishing effort by distant-water fleets (47) (Figure 3.7). Stock biomasses declined quickly and many fisheries had reached crisis status by 1974. Extended fisheries jurisdiction brought low groundfish quotas, a reduction of distant-water fleet effort, and enforcement of minimum mesh size restrictions, resulting in lower fishing mortalities for all age-groups and particularly for younger fish. Juvenile survival rates (48) increased and recruitment of young fish (mainly cod and haddock) improved, partly because of reduced exploitation in the fisheries and partly because of improved environmental conditions. Stock biomasses and landings increased in Canada and stopped declining in the United States.

Improving resource conditions and less competition from distant-water fleets spurred the development of domestic harvesting capacity and effort in the late 1970s and 1980s. Trawl fishing effort approximately doubled between 1976 and 1984 in the United States and some regions in Canada. Furthermore, vessels constructed after 1976 were generally larger, more technologically sophisticated, and better able to fish in difficult seas throughout the year (49).

The result was that groundfish stocks were overfished, primarily by coastal states, over the past 15 years. The United States had no explicit fishing mortality rate or biomass threshold objectives for groundfish during 1982–1993, and fishing mortality increased to very high levels. An evaluation of recent levels of F relative to spawner replacement targets indicates that most of the U.S. stocks are overfished (Figure 3.8). Canada had explicit $F_{0.1}$ management criteria but actual management performance was similar to that in the United States. For a variety of reasons, fishing

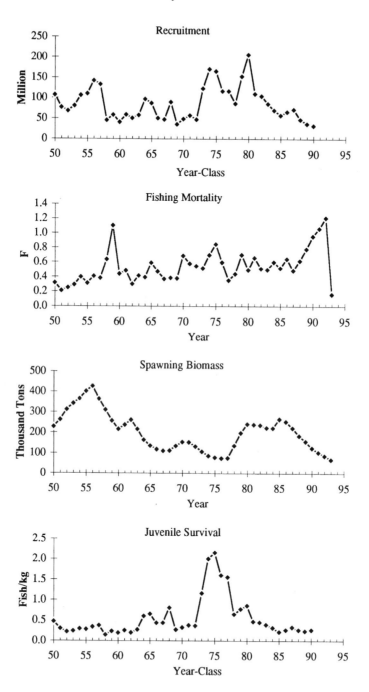

FIGURE 3.7.—Population and fishery trends for Atlantic cod in the southern Gulf of St. Lawrence, 1950 through 1990 or 1993. Recruitment is abundance at age 3, plotted at the year of birth ("year-class"). Fishing mortality (F) and spawning stock biomass are plotted at the year they were estimated. Juvenile survival is the number of age-3 recruits divided by spawning stock biomass in the year of birth, plotted at the birth year (year-class). The trends shown are similar to those for other cod stocks in the northwest Atlantic.

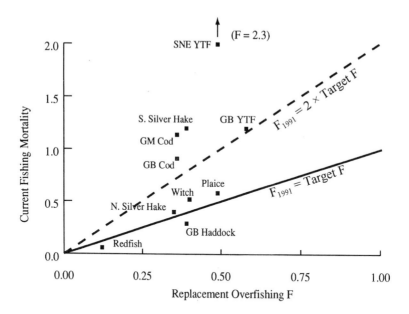

FIGURE 3.8.—Comparison of estimated fishing mortality rates ("current fishing mortality") with target fishing mortalities (F), designed to allow replacement of spring spawning stock biomass ("replacement overfishing"), for U.S. groundfish stocks in 1991. Only redfish and Georges Bank haddock were fished below their target levels (i.e., below the line F_{1991} = Target F); several stocks were fished at a rate more than twice the targets (i.e., above the line F_{1991} = 2 × Target F). Abbreviations: GB = Georges Bank; GM = Gulf of Maine; N and S = northern and southern stocks of silver hake; SNE = southern New England; Witch = witch flounder; YTF = yellowtail flounder.

mortality in Canadian groundfish stocks consistently exceeded reference levels (50) (Figure 3.9). The stock assessments were overly optimistic, largely due to unreported landings, discards, and misreporting of catches by area. The industry resisted reduced catch quotas. Quotas were not set low enough to achieve $F_{0.1}$ levels even when it became apparent that actual fishing mortality rates were consistently higher than predicted. To make matters worse, fishing mortality in several Canadian stocks increased dramatically in the late 1980s and early 1990s when stock biomass was falling. There is evidence that promising year-classes that should have supported the fisheries in the mid-1990s were substantially reduced because they were caught prematurely with illegal fishing gear and discarded at sea (51).

Several analyses in both countries have shown that harvest rates during the 1980s and early 1990s exceeded—often greatly—the rates at which spawning biomass could be sustainably replaced (52). Given such imbalances between harvest rates and production, substantial reductions in groundfish abundance were inevitable. Both countries practiced highly risk-prone management policies.

Signs of Overfishing

The increased landings of cod, haddock, and some flatfishes in the early 1980s masked several important warning signals that these harvests were not sustainable. Normal age structures were severely distorted as the older, larger

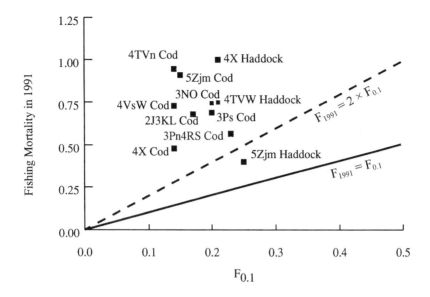

FIGURE 3.9.—Comparison of estimated fishing mortality rates with fishing mortality targets ($F_{0.1}$), determined by yield-per-recruit analysis, for Canadian groundfish stocks in 1991. Stocks are named by NAFO statistical conventions, except "5Zjm" refers to stocks jointly managed by Canada and the United States on Georges Bank. None of the stocks were fished below the target (i.e., below the line $F_{1991} = F_{0.1}$), and all but 5Zjm haddock were fished at a rate more than twice the target (i.e., above the line $F_{1991} = 2 \times F_{0.1}$).

fish were removed. The fisheries became highly dependent on newly recruited year-classes, which made up over 50% of the catch in some cases, and they became "recruitment fisheries." Much or most of the reproductive potential was lost from these populations (53). Although overall groundfish landings rose slightly during the 1980s, the increase was not general; several important stocks of haddock, redfish, and flounders collapsed.

Some of these changes were difficult to interpret when they occurred; in any case, management did not respond to them. Their mutual co-occurrence with a strong rise in fishing effort and fishing mortality made them viable indicators of troubled fisheries.

What if *F* Had Been Lower?

To what extent could risk-averse fishery management practices have avoided stock declines? The question has been approached in several ways. Computer simulation models, which describe a fishery mathematically and show the effects of altering various fishery properties and processes, have been developed for Atlantic cod stocks in Canada and the United States (54). Stock assessments provide a history of stock size, the abundance of young fish (recruitment), and fishing mortality. The basic approach in these simulations was to go back in time (usually to the 1970s) and calculate what might have happened under different conditions of fishing mortality and recruitment. In all simulations, whether recruitment was assumed to remain the same or to vary with stock size, markedly larger populations resulted when the target fishing mortalities (based either on yield per recruit or replacement spawning) were applied to the stocks. If the recruitment pattern from the most recent stock assessment was used, the simulated stocks declined in the 1990s, but much less than the real stocks. If the simulated recruitment varied with stock size, there was no decline in abundances in the 1990s.

These studies and other analyses of fishery dynamics already mentioned (55) offer compelling evidence that most of the important groundfish stocks in the NAFO area have been persistently overfished throughout the 1980s and early 1990s. Had management been able to keep fishing mortality below its actual levels and closer to biologically relevant targets, the fisheries would be productive today.

CONCLUSIONS

Groundfish landings, stock sizes, and production have declined considerably in the northwest Atlantic during the past four decades. Fishing mortalities were above sustainable levels for most of this time. Stocks were seriously depleted as distant-water fishing greatly expanded in the late 1960s and early 1970s. When Canada and the United States extended their fishing jurisdictions in 1977, distant water fishing greatly declined. For a few years, while the domestic fleets were growing, fishing mortality was reduced. Juvenile survival improved and so did recruitment, stock size, and landings. Increased effort and catches in the early 1980s masked early warning signs that harvest rates were approaching and exceeding sustainable levels. The loss of historically important stocks and spawning components of haddock, cod, redfish, and others, as well as changes in biological characteristics, were warnings of harvest-induced stress in the groundfish complex. During the 1990s, recruitment dropped below average in several important stocks. Fishing practices for much of the preceding decade had truncated the age compositions of stocks, reducing the spawning contributions of older fish, and creating "recruitment fisheries." Fishing mortality exceeded replacement levels in a majority of stocks, often by a factor of three or more. Declines in growth rates in some northern stocks and increased rates of fishing in the late 1980s and early 1990s exacerbated the situation. The ability of stocks to absorb and recover from additional stress, whether from environmental change or interspecies competition, was greatly reduced.

We conclude that the major reason for the decline of northwest Atlantic groundfish resources has been persistent recruitment overfishing. Although environmental variation likely has important effects on stock production, we found no environmental factor that could explain either the general decline in productivity observed since the 1950s or the precipitous decline in the 1990s. Little evidence suggests that changes in community structure or increased predation by marine mammals has caused these declines. The one common factor shared by almost all of the resources has been overfishing. Had recruitment overfishing been prevented, catastrophic declines in these resources could have been averted.

Rebuilding the groundfish resources of the northwest Atlantic will require a decade or more of highly restrictive management measures; fishing mortality rates will have to be kept at or near zero for many stocks. Southern stocks should recover more quickly than northern ones, owing to their shorter generation times.

Changes in the marine ecosystem, including increases in groundfish predators and competitors, may slow the rate at which the stocks increase. Because fishing must be sharply curtailed, monitoring the recovery will require increased reliance on research vessel surveys, conservative and carefully planned "sentinel" fisheries, and better recognition of basic life history characteristics. Along with improved monitoring must come a substantial change in management approaches. More emphasis should be placed on protecting the resource by implementation of risk-averse management policies and less on short-term economic and sociological concerns. The fate of these fisheries has provided a convincing test of the hypothesis of recruitment overfishing—a test that need not be repeated.

NOTES

1. Our information was compiled from journals, books, and many other stock assessment reports produced by government laboratories in Canada and the United States.
2. Pinhorn and Halliday (1990); Anthony (1993).
3. Frank et al. (1994).
4. Atlantic cod and other fish are aged by growth patterns in their scales and bones. These patterns record the slow growth during winter and more rapid growth during warmer seasons. For most fisheries (including those of the northwest Atlantic), fish are graduated to the next age-class on 1 January of each year, regardless of the month in which they were spawned. Thus a fish spawned in the fall will be assigned an age of 1 in January of the following year. This convention helps identify the year in which a fish was born and the "year-class" to which it belongs.
5. The decline in water temperature in this area is real (see Chapter 1 for details). The case for a temperature-driven shift in distribution and spawning was made by deYoung and Rose (1993) and Rose et al. (1994), but the case was disputed by Hutchings and Myers (1994b). In addition, scientists surveying juvenile cod in 1994 found evidence of spawning in division 2J (Anderson and Dalley 1995).
6. Sinclair (1996).
7. Millar and Myers (1990) found that the annual increases in weight at age of several cod stocks were correlated with both the total weight (or biomass) of the population and the temperature in the area. They were not able to identify whether population biomass or temperature was the more important factor.
8. Millar and Myers (1990); Frank et al. (1994).
9. Atlantic cod in the southern Gulf of St. Lawrence begin to reach a commercially acceptable size at age 4; however some of the slower-growing fish may not reach this size until age 6. Hanson and Chouinard (1992) suggested that the commercial fishery may selectively remove the faster-growing fish from an age-class. Because the remaining fish are smaller, the average weight at age of the age-class is deflated.
10. Anonymous (1995a).
11. Anonymous (1995a); Dutil et al. (1995)
12. The published material on trends in weight at length are derived from statistical relationships. No information is available on the variability of the indexes and it is difficult to judge if the reported changes are important. For example, the "low" values of weight at length of cod off eastern Nova Scotia in 1994 and of the condition of cod in divisions 2J and 3K in 1992 are about 10% below the values of the 1980s (Anonymous 1995d; Taggart et al. 1994), but whether these differences are meaningful is open to conjecture.
13. Lilly (1994, 1995).
14. O'Brien et al. (1993).

15. Morgan et al. (1994).
16. Mann (1993).
17. When an analysis deals with the abundance of young fish, the environmental data may be "lagged" by one or more years; that is, the status of fish now may be correlated with environmental conditions at some previous time in the cohort's history (e.g., in the year of hatching). The maximum possible lag depends on the age at which the fish enter the fishery.
18. Sutcliffe (1972, 1973); Sutcliffe et al. (1977).
19. Drinkwater (1987).
20. Colbourne (1995).
21. Sutcliffe et al. (1983).
22. Myers et al. (1993).
23. Hutchings and Myers (1994b) and Shelton and Atkinson (1994) reanalysed the salinity and recruitment series for Newfoundland cod and found that stock size, and not salinity, was the most important factor influencing recruitment.
24. Murawski (1993a).
25. Sutcliffe et al. (1983).
26. Drinkwater (1987).
27. Hutchings and Myers (1994b). See Chapter 1 for additional information on water temperature in the NAFO area.
28. Koslow (1984) and Koslow et al. (1987) did the initial analyses. Subsequent refinements were made by Thompson and Page (1989), who recognized the different patterns shown by Atlantic cod and haddock, and by Cohen et al. (1991) and Sinclair (1996), who provided north–south delineations.
29. Koslow et al. (1987) were the first to suggest that large-scale atmospheric pressure variations influenced cod and haddock recruitment. Variations in the NAO pressure index reflect passages of storm systems, which can influence mixing and movement of oceanic water masses in the northwest Atlantic and thus the amount of nutrients available to young fish and the ability of young fish to hold position in a dynamic current system.
30. Gomes et al. (1992).
31. Atkinson (1992); Anonymous (1995c); Gomes et al. (1995).
32. Mahon and Smith (1989).
33. Anonymous (1995d).
34. Gabriel (1992). Species diversity refers to the number of species and their relative abundances in an area. High diversity means there are many species and little difference in abundance among them. Low diversity means there are few species and one or two dominate in abundance.
35. Murawski (1991); Murawski and Idoine (1992); NEFSC (1995).
36. Rago et al. (1994); NEFSC (1995).
37. Overholtz et al. (1991); Northeast Fisheries Science Center (Woods Hole) file data.
38. Hammill and Mohn (1994) and Mohn and Bowen (1996) have studied grey seals; Stenson et al. (1995) and Shelton et al. (1995) have studied harp seals.
39. Overholtz et al. (1991).
40. Fishing mortality functions like a compound interest rate, except in reverse. For a given value of F, more fish are killed when the population is large than when it is small. For a given population size, a larger F means more mortality. Values of F can vary upward from 0.00 without limit, but they usually fall below 2.00 even for very heavily fished stocks. Fishing mortality can be calculated for a whole population or for identifiable parts of a population, such as size- or age-groups. In fishery analysis, all nonfishing sources of mortality (predation, starvation, disease, etc.) are lumped as "natural mortality." The sum of fishing and natural mortality gives "total" mortality.

41. Fishing mortality and fishing effort are thought to be proportional in fisheries—higher fishing mortality implies higher fishing effort, and therefore higher costs of fishing.
42. Fish populations vary in abundance from year to year. Fisheries biologists estimate population abundance in annual stock assessments. Using the predetermined fishing mortality target, they can also calculate the corresponding annual harvest.
43. Gulland and Boerema (1973); Pinhorn and Halliday (1990).
44. Anthony (1990, 1993).
45. Pierce (1982), cited by Anthony (1990).
46. Sissenwine and Shepherd (1987); Gabriel et al. (1989); Goodyear (1993).
47. Brown and Halliday (1983).
48. The term "juvenile survival rate" or "ratio" refers to the number of fish in a year-class that recruit to the commercial fishery divided by the spawning biomass that produced them. It is taken as an index of conditions faced by the year-class from the time of spawning to the time it reaches a commercial age. For the same spawning stock size, more recruits enter the commercial fishery when the survival ratio is high than when it is low. Many things may affect the juvenile survival rate, including the number of eggs produced and fertilized per unit spawning biomass, larval survival, and survival of fish between the larval stage and the time they enter the fishery. All these factors are integrated in (but not distinguished by) the juvenile survival ratio.
49. In Canada, these fishery developments have been documented by Pinhorn and Halliday (1990), Halliday et al. (1992), Kirby (1982), Hutchings and Myers (1994), O'Boyle and Zwanenburg (1994), and Sinclair et al. (1994). Trends in the United States have been recorded by Serchuk and Solowitz (1990), Mayo et al. (1992), Anthony (1993), and NEFSC (1995).
50. Sinclair et al. (1991); Maguire and Mace (1993); O'Boyle and Zwanenburg (1994); Sinclair (1994).
51. Sinclair et al. (1995); Myers et al. (1997).
52. Gabriel et al. (1989); Mace and Sissenwine (1993); Maguire and Mace (1993); Rivard and Maguire (1993); Shelton and Morgan (1994); Shelton (1995).
53. Trippel (1995).
54. Fréchet (1991), Chouinard and Fréchet (1994), and J.-J. Maguire (unpublished data) simulated the effects of lower levels of fishing mortality on stock size. In their simulations, they assumed that recruitment would not have changed in relation to the resulting increased stock size. We also investigated the effects of lower fishing mortality but assumed that the ratio between the number of recruits in a year-class and the spawning biomass that produced them (the juvenile survival rate) remained constant—that is, recruitment would change in relation to stock size. These results have not been published yet.
55. See especially Hutchings and Myers (1994b), Myers and Cadigan (1994), and Shelton (1995).

Ralph G. Halliday and Allenby T. Pinhorn

POLICY FRAMEWORKS

Commercial fishing is an industrial activity that people engage in to earn wealth and thus to maintain or enhance their standards of living. At a societal level, fisheries create employment, contribute to local and national economies, maintain the viability of coastal communities and the level of settlement in remote areas, and so on. The resource base, the fish, that supports this economic activity must be conserved if the economic and social aspirations of interested parties are to be met over the long term. A fisheries policy, which is simply a general plan of action adopted by government, could concern any aspect of fishing, and usually several—sometimes many—policies provide the overall policy framework within which a fishing industry operates.

Conservation policy is invariably viewed as the central element of a fisheries policy framework and, indeed, it is often equated with overall fishery management policy. However, conservation is fish stock management, whereas fishery management encompasses not only conservation but socioeconomic policies as well. Nonetheless, this chapter concentrates mainly on conservation policy because this constitutes the most developed (and best documented) policy element, and it provides the foundation on which social and economic aspirations are built.

Each policy or plan, if it is to provide adequate guidance for the rational management of a fishery, does several key things. It documents the reasons for its existence, the problems to be solved, and the opportunities to be capitalized upon; that is, it explains its objectives. The plan lays out the actions (strategies) needed to achieve these objectives and also the methods (tactics) by which these strategies are to be implemented and the ways in which compliance with these tactics is to be obtained (enforcement strategy). Underlying the plan is (or should be) a technical analysis demonstrating that the objectives will be realized if the specified tactics and strategies are implemented. Finally, the plan provides for its own evaluation by ensuring that appropriate data are collected and by establishing procedures to determine if the various elements of the plan are functioning effectively and the original objectives of the policy are being met.

96 Policy Frameworks

The following accounts of policy frameworks for the northwest Atlantic fisheries are ordered to address each of the planning elements: objectives, strategies, regulatory tactics, enforcement provisions, and measurements of compliance. Evaluation of fishery performance in relation to policy objectives is outside the scope of this chapter.

POLICY FRAMEWORKS

From 1950 until 1977, when coastal nations extended their jurisdiction seaward to 370 km (200 nautical miles), the regulatory authority for fisheries in the western part of the north Atlantic lay with the International Commission for the Northwest Atlantic Fisheries (ICNAF). Since 1977, authority for fisheries inside the 370-km limits has lain primarily with the United States and Canada. Important fisheries beyond 370 km on the Grand Bank and on Flemish Cap, southeast of Newfoundland, came under the regulatory authority of a new international commission, the Northwest Atlantic Fisheries Organization (NAFO). The policies of these four management authorities are considered here. Although west Greenland fisheries were also encompassed by the ICNAF Convention Area, post-1977 management in west Greenland waters is ignored for present purposes. The islands of St. Pierre and Miquelon, immediately south of Newfoundland and under the jurisdiction of France, are also ignored in this section.

International Commission for the Northwest Atlantic Fisheries

The first conservation policy for the northwest Atlantic groundfish stocks was developed within ICNAF, and this policy provided a common starting point for the management regimes of Canada and the United States after their extensions of jurisdiction. It is necessary, therefore, to be aware of ICNAF policy if the evolution of subsequent policies is to be understood.

The mandate of ICNAF was "the investigation, protection and conservation of the fisheries of the Northwest Atlantic Ocean, in order to make possible the maintenance of a maximum sustained catch from those fisheries" (1). Achievement of this objective of maximum sustained yield or catch (MSY—the largest average catch that can be taken from a stock on a continuing basis) was to be pursued through scientific investigations. A modification of the ICNAF Convention in 1971 changed its objective to optimal utilization of the stocks, rather than MSY. Although this amendment to the Convention broadened the basis for management to include economic and technical considerations, biological considerations remained the primary basis for regulation. However, the Commission usually continued to aim for MSY because, in its view, it represented optimum utilization.

The Convention initially restricted the Commission to proposing the following regulatory measures: (a) establishing open and closed seasons; (b) closure to fishing of spawning areas or areas populated by small or immature fish; (c) establishing size limits for any species; (d) prohibiting the use of certain fishing gears and appliances; and (e) prescribing an overall catch limit for any species of fish. The 1971 Convention amendment eliminated this list and gave the Commission scope simply to make "appropriate proposals" for regulations. The primary significance of this was to allow national allocation of overall (global) catch quotas, and this paved the way for acceptance of a comprehensive catch quota control scheme. It also permitted direct regulation of fishing effort.

In its initial years, the Commission dealt exclusively with regulatory requirements of the groundfish fisheries; in particular, it sought to prevent the wastage of small fish at sea. When the Commission was established, a need to regulate haddock fishing on Georges Bank was already recognized. The commonly used otter trawl mesh size was 73 mm, and large quantities of fish too small to market were being caught and discarded (2). A minimum mesh size of 114 mm was required in nets from 1953. Otter trawl mesh regulation was expanded to include Atlantic cod and extended to the Scotian Shelf, Gulf of St. Lawrence, and Grand Bank in 1957 and to Labrador and west Greenland in 1968. By the early 1970s, minimum mesh sizes had been increased to 130 mm (manila equivalent) in all areas and extended to various other groundfish species (3).

It was recognized in the 1960s that control of the size of fish caught was not enough to prevent overfishing. The international fleet was expanding its fishing effort steadily, and scientists were beginning to warn that the level of exploitation (fishing mortality) of some stocks was at or above the level associated with MSY. A 1965 scientific report to the Commission advised that "there must...be some direct control of the amount of fishing" (4). The first regulations intended to control fishing effort, and hence fishing mortality, were adopted for the haddock stocks off southwestern Nova Scotia and New England in 1969 for application in 1970. The method chosen was to set a total allowable catch (TAC) for each stock. Once the Commission acquired the authority to propose national allocation of TACs in 1971, catches from many other stocks were also regulated. By 1974, virtually all stocks in the northwest Atlantic that were subjects of a directed fishery were under TAC control. Off New England, an overall "second-tier" TAC was established in 1974, in addition to single stock (first-tier) TACs. This second-tier TAC was set at a level below the sum of the first-tier TACs to address mixed-fishery and bycatch problems and to allow for species interactions.

Regulation of trawl mesh size and TAC were the primary methods used by ICNAF to control the pattern and level of exploitation of fish stocks, but closed areas and seasons and limits on the amount of fishing were also adopted. Minimum fish size regulations were not considered necessary for groundfish species.

Area and seasonal closures were used by ICNAF to reduce bycatch problems in small-mesh fisheries, to protect spawning fish from disturbance, to reduce the level of fishing for particular stocks, and to reduce interference between fisheries. The Commission did not use closures for protecting small groundfish from capture, because small groundfish are not usually well segregated from commercial sizes and thus closures were thought not likely to be effective. The most important use of these closures was to address bycatch problems, which were most severe in the south of the convention area off Nova Scotia and New England. From 1974 onward, increasingly large portions of the shallow areas on and south of Georges Bank were closed to vessels engaged in small-mesh trawl fisheries. In 1977, these area closures were repealed in favor of "open-window" regulations, which defined the areas where, and seasons when, fishing for particular species could take place (the converse of closed areas and seasons). A similar measure restricted small-mesh fisheries to the edge of the Scotian Shelf. Although these windows were negotiated through ICNAF in 1976, they could be looked upon as national regulations because they applied to areas that came entirely under U.S. and Canadian jurisdiction in 1977. Closures of haddock spawning areas off southwestern Nova Scotia and New England from 1970 were ostensibly to improve spawning success, but they were part of a package of measures intended to reduce fishing mortality (5). Other uses of area and seasonal closures were of less importance.

Fishing effort regulation off New England was extensively discussed in ICNAF, but the proposals made proved too technically complex to gain acceptance, and the second-tier TAC was adopted instead. A simpler effort regulation was implemented off the Canadian coast in 1976; it required substantial reductions in the number of days fished by each vessel size category and gear type, but it lapsed a year later with extension of national jurisdiction.

It was up to individual members of ICNAF to ensure that their vessels conformed with the regulations in force. In addition, an ICNAF scheme for joint international enforcement, which became operative in 1971, provided for inspectors from each nation to board the vessels of other nations to verify that the Commission's regulations were being observed. In actuality, only Canada and the United States actively participated in at-sea inspections. Inspectors could only report apparent infractions to flag state authorities, and thus the scheme provided little in the way of deterrence. It did, however, establish that there was widespread disregard for ICNAF regulations.

Canada

Management authority for Canadian fisheries is vested in the federal Department of Fisheries and Oceans (DFO). The DFO conducted a fundamental review of marine fisheries management policy in the mid-1970s and published a comprehensive statement of policy in 1976 (6). The MSY objective of ICNAF was rejected in favor of "best-use," to be defined by the sum of net social benefits derived from the fisheries and associated industries. A low-exploitation strategy (corresponding to a fishing mortality of $F_{0.1}$ [7]) was adopted as an initial proxy for best use in the absence of any social and economic analyses. Total allowable catches were retained as the primary method of controlling the level of exploitation. Another important element of Canadian strategy was to balance the catching capacity of fleets with available resources. Limited entry licensing was introduced for large groundfish trawlers in 1973 and for smaller groundfish boats in 1976 (although entry continued to be allowed for certain types of small boats and in particular areas). Rules limiting the size of replacement vessels were an element of licensing policy. At the time Canada extended its jurisdiction, groundfish abundance was low and emphasis was placed on stock rehabilitation. Annual fishing plans allocated restrictive total allowable catches among interest groups and coordinated the deployment of fleets over the fishing grounds and operating season to promote full use of fleets and resources. Although groundfish stocks recovered soon after extension of jurisdiction, the domestic fleet also expanded, especially the small-boat sector where increase was encouraged as an element of social policy. As a result, the demands of the fleet continued to exceed available catch opportunities. By the early 1980s, it was recognized that all new entries to the fishery had to be prevented and that the practice of buying larger replacement vessels had to be more severely limited. By then, however, the licensed fleet capacity was much above that required to fully harvest the resource, and DFO was required not only to maintain but to increase its involvement in day-to-day control of the fishery.

An independent Task Force on Atlantic Fisheries was commissioned by the government in 1982 in response to financial problems in the industry. It proposed new objectives that were accepted by the government and subsequently guided DFO policy. These were, in order of priority, (a) sustained economic viability of the fishing industry, (b) maximization of employment at reasonable income levels, and (c) Canadianization of the fishery within the Canadian zone. (It was an underlying assumption of the task force that a sound resource conservation policy was already in effect.) The $F_{0.1}$ strategy, limited entry, and the annual planning process remained in place. However, as recommended by the task force, individual catch quota shares (IQs or, when transferable, ITQs), which were assigned to specific boats or to fishing enterprises (then called enterprise allocations), increasingly replaced allocations to fleet sectors that were defined on the basis of boat size and gear type. This was expected to reduce the motivation to invest in additional fleet capacity, and thereby to improve both the economic viability of the industry and control overfishing mortality.

When the stocks of cod and several other species collapsed in the early 1990s, the government had to provide large-scale financial support to avoid extreme social hardship. In 1992, another task force was established "on incomes and adjustment in the Atlantic fishery"; it was the first task force to have a primary purpose of addressing social policy. The "adjustment" required, according to this task force, was a 40 to 50% reduction in groundfish harvesting and processing capacity. The need for clear policy objectives that give explicit priority to ecological and commercial sustainability was also stressed. Implementation of the many recommendations is ongoing and the full effects of the task force report have yet to be determined.

Trawl regulations inherited from ICNAF were modified in 1982; use of 130-mm-mesh netting was required regardless of construction material (3), and most groundfish species were included in the regulation. The primary purpose of the changes was to simplify the regulations to improve the effectiveness of enforcement. In 1991, mesh size was further increased to 145 mm for the area off Nova Scotia and on Georges Bank (8). Introduction of minimum fish sizes in 1988 for Atlantic cod, haddock, pollock, and Atlantic halibut was a regulatory innovation. For cod, haddock, and pollock, the minimum size was 41 cm, which was increased to 43 cm in southwestern areas coincident with the 1991 increase in mesh size. Minimum size of Atlantic halibut was 81 cm. In 1993, the Icelandic concept of real-time closures was introduced along with a no-discarding rule. Closure of specific fishing grounds is required when at-sea observers estimate that undersized fish (defined as above) exceed 15% of the catch by number.

There are many season and area closures in Canadian fishery regulations but most relate to gear conflicts or indirectly to catch allocation issues. Closures related to conservation that had been established by ICNAF were retained but the only addition was a year-round closure of an area on the eastern Scotian Shelf, which was instituted in 1987 to protect juvenile haddock (ages 0–3).

Enforcement of regulations is the responsibility of DFO, which directly or through contracts conducts shore-based, at-sea, and airborne surveillance. The armed forces are sometimes called upon, particularly for enforcement relating to foreign vessels. The at-sea surveillance is carried out by a varied fleet of surface vessels and a corps of observers. Observer coverage of foreign fleets fishing in the Canadian zone is essentially 100%. Coverage of domestic vessels varies from 100% in a few high-profile fisheries to almost nil for small coastal vessels. There is virtually no published information on enforcement effectiveness, and thus the level of compliance with regulations is essentially unknown. Anecdotal information suggests that dumping of unwanted species and size categories (high-grading), misreporting of capture area and of species caught, and underreporting or nonreporting of quantities landed have been serious problems in many fisheries at times. The level of compliance with mesh regulations is not known either, but minimum fish size regulations were generally ignored, at least until the last few years.

United States

The Fishery Conservation and Management Act of 1976 (FCMA, subsequently renamed the Magnuson FCMA, or Magnuson Act for short) gave the U.S. government authority over a marine "fishery conservation zone" between coastal state waters, which extended outward 5.6 km (3 nautical miles), and an extended jurisdictional limit of 370 km. The Magnuson Act took effect in March 1977 and established the framework for management in the conservation zone (9). The primary federal authority under this act is exercised by the Secretary of Commerce through the National Marine Fisheries Service, an agency of the National Oceanic and Atmospheric Administration within the Department of Commerce. The act required the formation of regional fishery management councils. These councils prepare fishery management plans for the fisheries within their geographical areas of authority and submit these to the Secretary of Commerce for approval.

Two councils, the New England and Mid-Atlantic, have responsibilities for northwest Atlantic fisheries, but the major U.S. groundfish fisheries are the responsibility of the former, and its actions are described below. The New England Council consists of the states of Maine, New Hampshire, Massachusetts, Rhode Island, and Connecticut and has authority for fisheries off their coasts. This Council has 17 voting members, 5 of these being the state officials with responsibility for marine fishery management and another being the regional director of the federal National Marine Fisheries Service. The remaining 11 members are appointed by the Secretary of Commerce from lists, provided by state governors, of qualified individuals, defined as people who are knowledgeable or experienced with fisheries management, fish conservation, or recreational or commercial fishing in New England.

The Magnuson Act embodied a set of "national standards" for fishery conservation and management. Every council plan must meet these standards in the eyes of the Secretary of Commerce before it is approved. The first requirement of these standards is to prevent overfishing and to achieve "optimum yield." Optimum yield (OY) is defined as the amount of fish that (a) will provide the greatest overall benefit to the nation, with particular reference to food production and recreational opportunities, and that (b) is prescribed as such on the basis of maximum sustainable yield as modified by any relevant economic, social, or ecological factor. The second standard requires management plans to be based on the best scientific information available. Other standards state that individual stocks are to be managed as units throughout their range, interrelated stocks being managed as a unit or in close coordination; that allocation of fishing privileges, if necessary, is to be equitable, without discrimination among residents of different states, and such that no person or company acquires excessive shares; and that management measures are to promote efficient use, to be flexible, and to be cost-effective.

Various regulatory tools are available to the councils. They can require permits and levy fees for fishing in the fishery conservation zone. They can designate areas where, and seasons when, fishing is restricted or prohibited. Controls can be placed on the types and quantities of fishing vessels, fishing gear, and other equipment used, and it can be required that position locators and other devices that facilitate enforcement are carried. Total allowable catches, catch quotas, and limited access systems can be established. Management plans can build in relevant management measures of states adjacent to the conservation zone and, indeed, can prescribe any other measures and restrictions considered necessary.

The first plan for groundfish off New England, the Fishery Management Plan for Atlantic Groundfish, became effective in March 1977. This plan essentially maintained the controls on harvesting established under ICNAF, but it eliminated differentials in trawl mesh size regulations and it introduced minimum fish size regulations for Atlantic cod and haddock at 41 cm (10). However, only fishing for Atlantic cod, haddock, and yellowtail flounder was regulated under the plan. Although the status of Atlantic cod and haddock markedly improved in 1977, investment in new fleet capacity, which was not regulated, quickly put catch controls in difficulty and required a rapid escalation of regulatory restrictions. Discarding of fish and misreporting of catches became widespread, and the plan was replaced in March 1982.

A criticism of the first plan was that it did not identify objectives although, based on the actions taken, the strategies adopted seem to have been to fish at MSY for cod and $F_{0.1}$ for haddock. Its replacement, the Interim Fishery Management Plan for Atlantic Groundfish, was intended only as a stopgap measure until a more comprehensive plan could be developed. It too concerned only Atlantic cod, haddock, and yellowtail flounder, and its objectives were restricted to providing minimal safeguards for stock conservation while reliable data were acquired on normal fishing patterns and on the biological attributes of stocks. No controls were placed on catch, fishing effort, or participation. Optimum yield was defined as the amount of fish actually harvested by U.S. fishers in accordance with measures in the plan. Conservation objectives to maintain adequate spawning stock sizes and enhance the prospects for successful spawning were addressed through regulation of mesh and fish sizes and by closures of spawning areas. Only large-mesh nets could be used in the western Gulf of Maine and on Georges Bank west to Cape Cod. Exemptions for small-mesh nets could be obtained for specific fisheries in this large-mesh area, although under restrictive conditions. Large mesh was defined as 140 mm after 1982, and minimum fish size for commercially caught Atlantic cod and haddock was increased to 43 cm.

The interim plan was replaced by the Fishery Management Plan for the Northeast Multispecies Fishery in September 1986. The objectives (termed "basic goals") for this plan were to allow the multispecies fishery to operate with minimum regulatory intervention, but to prevent stocks from reaching abundance levels so low that their ability to replace themselves would be prejudiced. The primary management strategy (called an objective in the plan) was to control fishing mortality, primarily of juvenile (immature) fish, so that stock sizes would stay at or return to levels representing adequate spawning potential. The plan established three categories of stocks. Georges Bank haddock and Gulf of Maine redfish were to be managed for stock rebuilding. Atlantic cod, haddock, and various flounders in the Gulf of Maine; Atlantic cod, yellowtail, and other flounders on Georges Bank; and yellowtail and other flounders in southern New England were to be regulated for acceptable spawning potential. Other stocks required no specific regulatory action at that time. An "acceptable level" of spawning potential for each stock was defined in relation to the potential egg production in an unfished stock, which was termed "maximum spawning potential" or MSP (11). The Council adopted 20% of MSP as an acceptable level in general, but 30% of MSP was chosen for Georges Bank haddock and the "largest feasible value" for Gulf of Maine redfish. Optimum yield from the multispecies fishery was defined as the catches resulting from implementation of the management program, identical to OY in the Interim Plan. It was also a strategy of the plan to promote the collection of accurate fishery data for, among other things, evaluating plan effectiveness (12).

Regulation of fishing again depended exclusively on minimum fish size and mesh size limits and on closed areas. However, these provisions were strengthened in this new plan and in four subsequent amendments during 1987–1993. Areas in which only large-mesh nets could be used were extended, the minimum fish size regulations were applied to more species, and minimum size for Atlantic cod, haddock, and pollock was increased to 48 cm in 1987. The season when haddock spawning areas were closed was extended, and a temporary closure system to protect concentrations of small or spawning fish was introduced in 1990 (but this was not a real-time Icelandic-type closure system).

The Secretary of Commerce was not satisfied that the Council's Multispecies Plan adequately addressed conservation requirements, and the plan received subsequent and similar criticisms both from its own Technical Monitoring Group set up to monitor plan performance and from outside agencies. The most influential criticism, however, came from the Conservation Law Foundation and the Massachusetts Audubon Society which, in 1991, sued the National Marine Fisheries Service and Secretary of Commerce for failing to prevent overfishing of Atlantic cod, haddock, and yellowtail flounder. A court settlement required the Council to amend the plan in such a way as to eliminate the overfished condition of these species. This required a radical change in management strategy. In a fifth plan amendment, introduced in 1994, a moratorium was placed on new entry to the fishery and a scheme was designed to reduce fishing effort by 50% over a period of 5–7 years. It became mandatory to maintain and submit log records, to accept at-sea observers and, for certain vessels, to install an electronic vessel tracking system. Minimum mesh size was increased to 152 mm. Yet further restrictions were imposed, on an emergency basis, at the end of 1994 that extended the area closures, increased the requirements to use large-mesh trawls, and reduced bycatch allowances. The seventh amendment, in mid-1996, brought further constraints on fishing effort and catches as attention turned to stock rebuilding.

As required under the Magnuson Act, the Coast Guard, which is part of the Department of Transportation, provides at-sea surveillance from ships and aircraft, whereas the National Marine Fisheries Service concentrates on shore-based enforcement. Council plans have applied to all U.S. waters and thus full implementation requires state governments to cooperate with complementary legislation and enforcement for the territorial waters in which they exercise jurisdiction. In the initial years of groundfish management, failure to integrate state and federal regulations left wide loopholes for circumvention of Council measures. From the beginning of Council activities in 1977, the fishing industry showed little acceptance of the need for direct control of fishing mortality, be it through catch limitations or otherwise. This resulted in the adoption of the principle of minimum interference embodied in plans from 1982. The Council anticipated that, because fishers appeared to want this approach, they would be willing to comply with those few regulations in the plan. The Council also expected that enforcement agencies had the ability to enforce these rules. These expectations were not met. One study showed that in the Georges Bank fishery, regulations were frequently violated by a quarter to a half of all fishers. These violators used illegal mesh on almost all trips and fished in closed areas on about one third of their trips. The Council's Technical Monitoring Group pointed out that there were few incentives for fishers to comply with regulations. There were inadequate resources for enforcement, and the plan contained regulations that were difficult to enforce and provided ready loopholes for evasion, so the risk of detection was low (13).

Northwest Atlantic Fisheries Organization

The NAFO Convention came into force in 1979 but ICNAF had continued to function in modified form since the 1977 extensions of national jurisdiction, so there was an uninterrupted and orderly transition of authority for management of those fisheries remaining outside the control of coastal nations. The ICNAF regulatory framework and enforcement provisions were transferred to NAFO in large part (14). The NAFO convention area was identical to the ICNAF statistical area, incorporating all waters north of Cape Hatteras and west of the southern tip of Greenland. Canada was a signatory to the NAFO convention but the United States, which had been a member of ICNAF, elected not to join the new organization until late 1995.

The objective of the NAFO Convention is to contribute through consultation and cooperation to the optimum use, rational management, and conservation of the fishery resources of the convention area. The Convention's regulatory responsibilities are discharged by a Fisheries Commission. This Commission's pursuit of optimum use is constrained only by a requirement to seek consistency with national regulatory measures when stocks are fished both in national zones and in the NAFO regulatory area (that part of the convention area outside 370 km). In practical terms, this constraint has applied to Grand Bank stocks exploited on both sides of the Canadian limit. Flemish Cap stocks of Atlantic cod, redfish, and American plaice are recognized as lying entirely within the regulatory area. Canada and ICNAF had agreed in 1978 that five groundfish stocks were of shared interest. This list was inherited by NAFO but, as more knowledge became available on stock distributions, the list at times became contentious and some changes occurred.

In 1976, ICNAF had adopted an exploitation strategy of $F_{0.1}$ as the basis for recommending total allowable catches to member governments, not only for most of the stocks that would be exclusively within domestic waters in 1977 but also for transboundary stocks and those outside 370 km. When NAFO took over from ICNAF, the use of $F_{0.1}$ continued. This met the requirement for consistency with the exploitation strategy used in the adjacent Canadian zone. However, the $F_{0.1}$ strategy came under serious challenge from the European Union (then the European Community) in 1985. The European Union believed that the socioeconomic needs of its fleet, about to be greatly augmented by the accession of Spain and Portugal, would be best served by fishing at F_{MAX}. Although NAFO retained $F_{0.1}$ as a basis for setting total allowable catches at Canadian insistence, the European Union systematically used the organization's objection procedure. This exempted it from legal obligations to comply and allowed it to establish much higher catch levels for itself under domestic legislation. A Canadian–European Union fisheries agreement reached at the end of 1992 committed the European Union to respect all NAFO decisions, but the agreement does not necessarily end the debate about management strategy.

The ICNAF trawl regulations were continued unchanged in NAFO regulations. The NAFO Fisheries Commission did not agree initially with Canadian proposals to dispense with differentials based on netting material and thus to obtain consistency with regulations established in the Canadian zone in 1982. However, increasing evidence, from 1989 onward, that substantial quantities of small fish were being caught changed that view, and minimum fish size regulations were instituted in 1992 at 41 cm for Atlantic cod and 25 cm for American plaice and yellowtail flounder. Then, in 1994, netting differentials were removed, all groundfish species were included in the mesh regulations, and the Grand Bank was established as a large-mesh-only zone.

Canada was the only active participant in the Joint Enforcement Scheme of ICNAF in the Grand Bank–Flemish Cap area, carrying out regular air and surface surveillance and boardings for inspection (15). Canadian proposals for an equitable sharing of the burden at the beginning of NAFO produced a response from the USSR but little from other members until the European Union began deploying inspection vessels in the regulatory area in 1988 and a Faroese inspection vessel participated in 1989. The enforcement scheme was modified and strengthened over time, the most significant change perhaps being institution of an experimental observer scheme in 1993. However, in NAFO, as it was in ICNAF, ensuring that vessels conform to NAFO regulations is the responsibility of the members to which the vessels belong. The multilateral scheme provides only for coordination of enforcement efforts and some third-party verification of regulatory compliance.

Noncompliance with the NAFO management program has presented a serious threat to the organization's conservation objectives throughout its history. The systematic objection of the European Union to catch quotas set by NAFO and the Union's unilateral establishment of much higher quotas for its fishers in 1986–1992, has already been mentioned. Noncompliance also took the form of a disregard by members of their legal obligation to conform to their quotas as agreed to within NAFO. European Union fishers, furthermore, at times exceeded the higher quotas established domestically. Yet another problem, equally serious in scale, was posed by the completely unregulated fishing in the regulatory area by nonmembers of NAFO (16). This disregard for NAFO regulations was not limited to catch quotas; others such as trawl mesh size requirements were also widely ignored.

With regard to the issue of nonmember fishing, the 1982 Law of the Sea Convention established the rights of fishers of all nations to engage in high-seas fishing. With these rights came obligations to cooperate with other states in the conservation of resources and, in the case of transboundary stocks, to respect the rights and interests of coastal states. The rights to fish in international waters have been well exercised, but the obligations have generally been neglected by third parties in the NAFO regulatory area. The problems this created in the northwest Atlantic and in other high-seas fisheries stimulated a United Nations Conference on Straddling and Highly Migratory Fish Stocks in 1992. The multiyear conference adopted an agreement for implementation of the 1982 Law of the Sea Convention with regard to these stocks at its sixth session in August 1995. The agreement opened for signature in December 1995 and comes into force when ratified by 30 nations. Its provisions will greatly strengthen the abilities of regional fisheries organizations to enforce their regulations. However, the Canadian government decided that more immediate action was required to address NAFO enforcement problems and, through amendments to its domestic Coastal Fisheries Protection Act, gave itself the authority to unilaterally enforce NAFO regulations for transboundary stocks in international waters. Under threat of arrest, nonmember vessels departed from the Grand Bank in mid-1994. This effectively protected Canada's immediate interests, but because these vessels redirected their fishing effort to the entirely international stocks on Flemish Cap, it did not solve all of NAFO's problems with nonmember fishing.

Canada confronted the lack of compliance with NAFO regulations by members in a like, but more dramatic, fashion in 1995 by arresting a European Union vessel in international waters for violation of NAFO-set Greenland halibut quotas (17). This precipitated serious diplomatic and at-sea confrontations which, nonetheless, resulted in agreement between Canada and the European Union to greatly strengthen enforcement measures, the most important of which is likely to be the requirement for 100% observer coverage on vessels fishing in the NAFO regulatory area. These enforcement measures were subsequently adopted by NAFO and thus became applicable to all members.

The weakness of the NAFO surveillance and enforcement scheme prevented NAFO from establishing effective control over exploitation in its regulatory area, and this in turn prevented successful implementation of its management strategies. More time is required to evaluate the extent to which anticipated benefits from recent improvements can be realized.

POLICY CONFLICTS

Whenever political boundary lines are drawn on a map of the oceans, transboundary resource management issues are created. When transboundary stocks are important to the fisheries of adjacent political regimes, effective conservation, and hence success in meeting the socioeconomic objectives of both parties, requires the parties to adopt consistent conservation policies. As already noted, although the NAFO Convention required consistency of approach between NAFO and Canadian conservation policies, this did not prevent a conflict between two of the most influential NAFO members, Canada and the European Union, over the NAFO strategy regarding exploitation levels, and this had serious adverse effects on the organization's effectiveness. The conflict extended to regulatory tactics and enforcement elements of policy. Other policy conflicts arose between Canada and France and between Canada and the United States.

The basis for policy conflicts between Canada and France was created in 1977, when Canada claimed jurisdiction of all waters beyond 22 km (12 nautical miles) from the French islands of St. Pierre and Miquelon, which lie immediately south of Newfoundland. France claimed a 370-km boundary or, where appropriate, a boundary equidistant between these islands and the Canadian coast. The result was a large disputed zone, the chief feature of which was St. Pierre Bank, a large productive area south of the islands. In the subsequent 5 years, agreements were reached on total allowable catches and national allocations for St. Pierre Bank cod, the primary resource in dispute, based essentially on a Canadian management strategy. Thereafter France acted unilaterally, which had the overall effect of greatly increasing the exploitation level of this stock. This precipitated a prolonged period of intense dispute and progressive deterioration in fisheries relations between the two parties, which ended only with referral of the boundary question to an international judicial tribunal. The tribunal's 1992 decision gave France only a narrow zone of authority (18), which left management of St. Pierre Bank cod essentially in Canadian hands (but the locally important Icelandic scallop fishery lies largely in French waters).

Canadian and U.S. jurisdictional claims on Georges Bank overlapped, but an interim agreement between the two parties controlled catch levels while a long-term fisheries agreement was being negotiated. However, these negotiations failed and the jurisdictional issue was referred to a chamber of the International Court of Justice in the Hague (19). In 1984, the Court awarded the northeast corner of Georges Bank to Canada and the remainder to the United States and, as a result, caused several groundfish stocks on the bank, most importantly those of Atlantic cod, haddock, and yellowtail flounder, to become transboundary resources. Resolution of the boundary issue thus defined a requirement for consistent management of these species on either side of the Georges Bank boundary. However, the radically different national approaches to management policy that had developed in the interim made agreement on a joint approach seem dim, and this deterred negotiations. Thus, each party managed its own stock components without reference to the actions of the other. The first practical recognition of a need for consistent conservation actions occurred in 1994, when Canadian and U.S. management authorities agreed to strengthen protection of the Georges Bank haddock stock, which had reached a very low level by the end of 1993.

OVERVIEW

From a common heritage in ICNAF, Canadian and U.S. groundfish management policies diverged radically after 1977, when both countries established extended jurisdictions. The Canadian low-exploitation strategy of $F_{0.1}$, combined with a policy of high levels of employment in the fishery, could be achieved only through large-scale government intervention to control the fishing activities of participants. The United States took the opposite approach of minimizing regulatory interference and not controlling exploitation level. The outcomes of both national policies have been judged unsatisfactory, and each regime is in the process of implementing substantial changes. For the area outside 370 km, NAFO has spent much of its existence in conflict over strategy, and it could perhaps best be described as a regime verging on anarchy. Disputes about policy among adjacent regimes, and among members within NAFO, have been the norm rather than the exception. Central to these differences of view was the Canadian fidelity to a low exploitation strategy but, of course, many other factors were also relevant.

The management of a fishery is not a controlled experiment and it is a complex task to separate the role of policy from the many factors determining the course of fishery events. It is clear, however, that the simple articulation of policy, without subsequent actions to implement and enforce it, cannot be expected to greatly influence conduct of a fishery. Although data on compliance are sparse, there is enough evidence to conclude that no regime has paid sufficient attention to effective implementation of conservation measures.

The standard enforcement model is that deterrence against violations increases with the probability of being detected and the severity of penalties. When fishers are highly motivated to circumvent the regulatory regime, the costs of enforcement may be higher than a management authority is able to pay or can justify paying given the economic value of the fishery. Furthermore, the severity of penalties may have to be greater than society is prepared to accept as reasonable for this category of crime. Only recently has serious attention been directed at the factors that motivate fishers to violate regulations (see Chapter 5). Regulatory mechanisms are required that modify fishers's behavior by creating incentives for compliance. This will not remove the need for enforcement of regulations, but the

108   Policy Frameworks

resultant reduction in violations could make enforcement against remaining violators more tractable and thus bring compliance to a high enough level that policies can be successfully implemented.

NOTES

1. The text of the International Convention for the Northwest Atlantic Fisheries is provided in the first (1951) of a series of annual reports and proceedings that document ICNAF regulatory actions. These publications are available from the Executive Secretary, Northwest Atlantic Fisheries Organization, Post Office Box 638, Dartmouth, Nova Scotia B2Y 3Y9.
2. Graham (1952).
3. Trawl mesh regulation was first introduced when manila was in widespread use as a netting material. Synthetic materials quickly replaced natural fibers in net construction in the 1950s. Nets of these materials retained a different proportion of small fish (usually less) than manila nets with the same mesh size. However, the selection characteristics of manila were retained as the standard against which other materials (natural and synthetic) were rated. These ratings were expressed in regulations as "differentials" from manila and allowed nets with smaller mesh sizes to be used if netting material was cotton, hemp, polyamide, or polyester. A differential based on gear type was allowed for Danish seine nets. Thus, a regulation requiring use of 130-mm mesh, manila equivalent, would allow trawlers to use a mesh of 120 mm and Danish seiners a mesh of 110 mm in most cases. See Holden (1971).
4. Templeman and Gulland (1965).
5. These spawning season closures were proposed by fishers in the belief that protection of fish while they are in the act of spawning has direct biological benefits. In the absence of proof, biologists tended to discount this possibility but they, and regulators, recognized the potential of these closures, in conjunction with TAC regulation, to aid control of fishing by spreading the fishery more evenly throughout the year (Halliday 1988).
6. The three key documents concerning Canadian policy for the Atlantic groundfish fishery are FMS (1976), Kirby (1982), and TFIAAF (1993). Parsons (1993) wrote a comprehensive policy review. Further information is available from the Communications Directorate, Department of Fisheries and Oceans, 200 Kent Street, Ottawa, Ontario KIA OE6.
7. In ICNAF, F_{MAX}—the fishing mortality giving maximum yield per recruit—was commonly taken as a proxy for MSY. A new biological reference point was introduced in the early 1970s, $F_{0.1}$ (Gulland and Boerema 1973), a level of fishing mortality about one third lower than F_{MAX}. It is defined as the fishing mortality at which the marginal yield (the increase in yield achieved by adding another unit of fishing effort) is 10% of the marginal yield when the fishery is very lightly exploited.
8. Nets with square mesh in the cod end are allowed a mesh size of 130 mm (with some exceptions), because this shape and size selects the same sizes of roundfish as a traditional diamond mesh cod end of 145 mm.
9. The Magnuson Act was amended and reauthorized, on 11 October 1996, as the Magnuson-Stevens Fishery Conservation and Management Act.
10. The minimum size was actually 16 inches (40.6 cm) but sizes, in this and other cases, are rounded to the nearest centimeter.
11. Sissenwine and Shepherd (1987) explained the theory underlying calculations of spawning potential. Spawning stock biomass is another measure of spawning potential.

12. Fishery management plan regulations are published in the U.S. *Federal Register*. Information on these plans and supporting documentation is available from the New England Fishery Management Council, Suntaug Office Park, Saugus, Massachusetts 01906-1097. Anthony (1990) reviewed groundfish management in New England.
13. Sutinen et al. (1990) gave measures of noncompliance. A report entitled "An Assessment of the Effectiveness of the Northeast Multispecies FMP [Fisheries Management Plan] with Recommendations for Plan and Management System Improvements" was provided to the New England Fishery Management Council by its Technical Monitoring Group, dated 22 June 1988.
14. The text of the Convention on Future Multilateral Cooperation in the Northwest Atlantic Fisheries is provided in the first annual report (for 1979–1980) of the Northwest Atlantic Fisheries Organization established under it. Documentation of NAFO regulatory actions has been published in annual reports (1979–1980, and 1991–present) and meeting proceedings (1979, 1991–present). A summary of decisions during 1979–1992, published in 1993, compensated in some part for failure to publish reports and proceedings in the 1980s. These and other NAFO publications are available from the Executive Secretary, Northwest Atlantic Fisheries Organization, Post Office Box 638, Dartmouth, Nova Scotia B2Y 3Y9.
15. The NAFO Convention specifically required that the ICNAF enforcement scheme be maintained in force. This was codified into NAFO regulation as the Scheme of Joint International Enforcement in 1981. A new scheme, which introduced only minor changes but with a symbolic revision in name, the Scheme of Joint International Inspection, was adopted in 1988.
16. Nonmember fishing can be divided into that by vessels operating as part of the domestic fleet of their flag nation and that by vessels flying flags of convenience. In the NAFO regulatory area, nonmembers of the first category were primarily Spain, until that country joined NAFO in 1983, the Republic of Korea, until it joined in 1993, and the United States, until it joined in 1995. In the second category, the many flags of convenience seen since 1978 have most frequently been flown by vessels of Spanish, Portuguese, and Korean origin.
17. Day (1995) provided an account of Canadian actions in 1994–1995 to unilaterally enforce NAFO regulations on the Grand Bank.
18. The tribunal decision gave France a 44-km zone southwest of St. Pierre and Miquelon and an approximately 19-km-wide corridor running due south of the islands for a distance of 370 km.
19. An Agreement on East Coast Fishery Resources was signed by the United States and Canada on 29 March 1979, but it failed to obtain the necessary approval of the U.S. Senate and was withdrawn from consideration on 6 March 1981, by President Reagan. Prior to this, an Interim Reciprocal Fisheries Agreement had been signed on 24 February 1977. A renewal for 1978 never entered into force and the agreement was abandoned by Canada in June 1978. Thereafter, both parties fished the disputed zone under independent domestic regulation until the boundary resolution of 1984.

Madeleine Hall-Arber and A. Christopher Finlayson

ROLE OF LOCAL INSTITUTIONS IN GROUNDFISH POLICY

A stroll down the pier of almost any commercial fishing port in the northeastern United States and Atlantic Canada will reveal to the most casual observer that fishing can be done with a broad array of vessels and gear. What will not be immediately apparent is that the style of fishing, attitudes towards conservation, world views, lifestyle, and community life in the fishing community may vary as much as the gear and vessels.

There seems to be little dispute among fisheries managers, scientists, and fishermen themselves in both Canada and the United States that the goal of fisheries management should be sustainable harvests that assure societies the greatest long-term benefit from the fishery (1). At the same time, fishing people in both countries embrace a strong egalitarian ethos that supports open access to fisheries resources. Managers of northwestern Atlantic fisheries have been unable to achieve these inherently conflicting goals thanks to the growth in fishing capacity, a world market that drives demand, and the diversity among fishing communities.

In this chapter, we focus on the effect of community institutions on national and regional fishery management policy. Ostrom defined institutions as "sets of working rules that are used to determine who is eligible to make decisions in some arena, what actions are allowed or constrained, what aggregation rules will be used, what procedures will be followed, what information must or must not be provided, and what payoffs will be assigned to individuals dependent on their actions" (2). Furthermore, the "working rules are common knowledge [to the participants] and are monitored and enforced." However, "working rules may or may not closely resemble the formal laws that are expressed in legislation, administrative regulations, and court decisions." Rules can remain informal and covert. In fact, working rules may even confer on individuals or other groups rights and duties that actually contradict rights and duties associated with their government or formal legal system.

The working rules that constitute institutions strongly influence individuals' actions. Ostrom noted that rational action depends on an individual's choice, based on "expected benefits, expected costs, internal norms and discount rates," but these in turn are affected by norms shared with others and the range of opportunities available. If one can learn the set of rules shaping the cost-benefit calculus of the members of a social group, one can make some predictions about how that group will respond to change. The job is complicated, however, because in a complex society, local rules often conflict with or change the meaning of regional and national rules.

Studies of small-scale, renewable resource management systems show that the inherent conflict between resource sustainability and open access is overcome at a local level by limiting use rights in ways that are consistent with the norms of local institutions (3). The community enforces rules for access and use because there is a clear set of stakeholders with shared interests. Several studies have pointed out that such management systems are most effective for homogeneous groups not subject to intense competition from outsiders (4). In these systems, fishery use rules also reflect the norms and practices in the community.

The problem with a federal management system, as we see it, is that a federal system encompasses many local communities and stakeholders with diverse interests. Members of a local community have little incentive to exercise social control over their own use of resources when the resources are accessible to people from other communities. Indeed, the incentive is to help local fishermen take more fish than outsiders do.

Equally important, conflicting interests become moral issues when national institutions differ from local institutions. Local norms about what is just and good may conflict with federal rules that determine access to federal fisheries or other rules that affect the distribution of resources. Furthermore, if the national process for public participation in management differs from local norms and practices, the federal process will lack local credibility. The history of the Atlantic groundfish crisis on both sides of the Canada–United States border provides examples of the power of local institutions to foster resistance to national programs.

In this chapter we examine the relationship between local institutions and federal management in order to elucidate the social and cultural dimensions of the current crisis in the northwest Atlantic groundfish fisheries. First for New England and then for Atlantic Canada, we briefly describe the fishery context and then use a community case study to illustrate the effect of local institutions on social responses to fishery management.

NEW ENGLAND

The story of New England's groundfish fishery began over 350 years ago when rival enterprises, the Plymouth Company and the Dorchester Company, each sent fishermen, a ship's carpenter, and a saltmaker to Gloucester, Massachusetts, to be close to the rich fishing grounds of Georges Bank. Thus began the lucrative trade in salt cod plied by picturesque, sail-borne vessels manned by proud, brave Gloucester fishermen—a romantic image enhanced by song, story, and statue.

Georges Bank, with its nutrient-rich upwellings, shallow water, and currents, provides an unusually productive habitat for groundfish. For many years, cod dominated the commercial fisheries landings. In the early Nineteenth Century, a great halibut fishery boomed—but went bust by the turn of that century. Haddock became popular when salt gave way to ice for preserving fish. Over time, yellowtail flounder and scallops joined cod and haddock to be numbered among the species most highly valued in the U.S. market.

As the United States developed, so did the fishing industry. The Gulf of Maine became renowned for its fish stocks and the coasts of Maine, New Hampshire, Massachusetts, Rhode Island, and Connecticut became lined with small fishing villages. Eventually, major ports and urban centers developed in Portland, Maine; Portsmouth, New Hampshire; and Gloucester, Boston, and New Bedford, Massachusetts. Newport and Point Judith, Rhode Island, and Stonington, Connecticut, also developed into important fishing ports.

But today, Georges Bank cod, haddock, and yellowtail flounder are believed to be on the verge of collapse. Overfishing has been identified as the principal cause of the stocks' depletion (Chapter 3). In consequence, the New England Fishery Management Council amended its Multispecies Fishery Management Plan, closing major portions of Georges Bank to commercial fishing and severely limiting the numbers of days fishermen can be at sea, among other restrictions. Scientists estimate that it will take 10 years for the cod and yellowtail flounder stocks to recover and 15 years for haddock even if there is no directed groundfish fishery on Georges Bank. Fishermen, their families, and the communities that have based their livelihoods on the fishing industry face an uncertain future with great trepidation. How did this happen? Why did the fisheries management system fail to restrict fishing to a sustainable level?

Fisheries Management

In New England, fisheries management may be said to have started in 1639, when the Massachusetts Bay Colony ruled that neither striped bass nor cod could be used as fertilizer (5). The first federal response to fisheries management issues was in 1871, when Congress created the U.S. Fish Commission to investigate the diminishing stocks of nearshore fisheries (6). In the 1930s haddock became the first of the Georges Bank groundfish (since halibut) to show signs of overfishing. When the International Commission for the Northwest Atlantic Fisheries (ICNAF) was created by Atlantic fishing nations, haddock was the subject of its earliest actions (7).

Despite ICNAF's efforts, distant-water fleets from the Soviet Union, Poland, Germany, Spain, and Japan moved onto Georges Bank to reap great fish harvests in the 1960s. The intense fishing effort from factory trawlers with their huge nets and accurate fish-finding techniques led to severe, unsustainable pressure on the stocks. New England fishermen were horrified by the enormous harvests, especially as contrasted with their own diminishing catches. In the early 1970s, a sharp downturn in the stocks of cod, haddock, and flounder led to an outcry that culminated in federal passage of the Fisheries Conservation and Management Act of 1976, commonly referred to as the FCMA or the Magnuson Act (for its principal sponsor) or, as it was reauthorized, the MFCMA.

With the Magnuson Act, U.S. sovereignty was extended from the traditional 12 to 200 nautical miles (from 22 to 370 km). At the same time, eight regional fisheries management councils were established and given responsibility for (in the words of the act) "conservation and management measures [to] prevent overfishing while achieving optimum yield from each fishery on a continuing basis." Managers were directed to modify their measures "by any relevant economic, social, or ecological factors." In other words, the objective of the councils is to establish fisheries management plans that balance ecological, economic, and sociocultural considerations so that the country derives the greatest benefit from the exploitation of these natural resources. The Secretary of Commerce reviews the plans to insure consistency with national standards and other laws and to suggest amendments and promulgate regulations. Enforcement is the joint responsibility of the National Marine Fisheries Service (NMFS) and the U.S. Coast Guard.

The New England Fishery Management Council consists of the heads of the 5 coastal states' marine fisheries departments, the regional director of NMFS, and 11 at-large members appointed by the Secretary of Commerce from nominees submitted by the state governors. All the regional councils have professional staff members who work with their councils and NMFS scientists to help design the management plans. The councils have independent scientific and industry advisory committees that are supposed to contribute to the development of the plans and to review proposed alternatives. The New England Council has been criticized by industry members and some scientists for not convening their advisory panels frequently enough.

When the Magnuson Act came into effect, the U.S. fleet was considered underdeveloped. Although it was recognized that the stocks had suffered under the pressure from foreign fishing fleets, most people believed that they would quickly rebound once those fleets were banned. Consequently, the U.S. government enacted various programs to help fishermen invest in the latest advances in new vessels and technology. Two programs—the Fishermen's Capital Construction Fund and the American Fisheries Promotion Act of 1980—have been most often blamed for having inappropriately encouraged and funded overcapitalization of the U.S. fishing fleet.

Inshore fisheries of New England—those within 3 nautical miles (5.6 km) of shore—are separately managed by the five coastal states. However, because the stocks are not stationary, the condition of the offshore stocks affects the catches of inshore fishermen. In addition, human-induced alterations of inshore habitats (e.g., estuaries and nursery areas) are thought to affect the status of both inshore and offshore fish stocks.

Diversity of Fish and Technology

One problem with the fisheries in New England is ecological. All 10 of the species covered by the New England Council's Multispecies Management Plan mingle near the bottom. Numerous other species also are found in the vicinity.

Not all of these fish are the same shape nor do they mature at the same size, so they are affected differently by various gears. A gear that catches roundfish only when they are mature, for example, may have a devastating effect on flatfish by catching juveniles. Furthermore, not all species are equally desired by the U.S. market. Nevertheless, they are all caught simultaneously by many of the gears used in the fishery today. Juveniles and those species underappreciated by the market are discarded.

Technological innovation has played an important role in development and change in the New England fisheries. In the early days of sail, schooners carried crews of 24 men who fished off dories using baited hooks on handlines. By the late Nineteenth Century, individual fishermen employed multiple tub trawls, each with 10 lines of 50 fathoms (91 m) and 55 hooks (i.e., 550 hooks per tub) (8). It was in this period, before technology had greatly changed, that halibut was fished to commercial extinction.

Use of steam engines in fishing vessels led to the development of the otter trawl fishery, which began in the United States in 1905. At that time, many fishermen feared that the otter trawl would lead to the loss of aspects of fishing that they valued. Though dory fishermen criticized the loss of independence, loss of traditional skills, and reduction in the labor force, and though they feared the effects of otter trawls on the abundances of key fish species, the trawl eventually became the dominant gear on Georges Bank (9). Nonetheless, handlining, longlining, and gillnetting remain options for fishermen. Groundfish are also caught by jigging, in weirs and other traps, and as a bycatch in shrimp trawls, scallop dredges and, to a lesser extent, lobster traps.

More recently, technological changes have improved fishermen's efficiency and thus contributed to overfishing. These innovations have occurred in navigational equipment that permit fishermen to return with extreme accuracy to the same spots they have fished before, in scanners and sonar that depict the bottom and identify schools of fish, and in rock hopper or roller gear that allows the otter trawls to be dragged over rocky bottom without ruining the nets. In the current crisis, the majority of fishermen cling to their own gear as being the "best" for the resources, and these fishermen accuse adherents of other gear of following various unsound ecological practices.

Today's groundfish vessels range in size from 30 to more than 100 feet (9–30+ m). Captains of smaller vessels tend to be opportunistic, shifting gears readily to capitalize on species availability and price changes. Sometimes the shifting is seasonal, but sometimes a skipper may switch from a large-mesh species to a small-mesh species "to top off the trip." Vessel sizes correspond to some extent with port size; that is, the smaller vessels tend to cluster in the smaller ports. Furthermore, small vessels tend to fish closer inshore than do larger boats.

The harvesting sector of the fishing industry spawns a wide range of support activities and related industries. Fishing has always required gear for catching fish and outfitting the vessels, food for the fishermen, salt and (later) ice for preserving the catch, fuel, a dock, a processing facility, a transportation system for moving the captured fish to market, and wholesale and retail markets. The support industries that grew up around the fishing industry created linkages throughout the community, the region, and (via exports) the world. Issues relating to the fishing industry therefore have ramifications that extend far beyond the fate of individual fishermen.

The institutional arrangements for fulfilling the needs of the fishing industry both affect and are affected by fisheries management. The global market offers opportunities for sale of both juveniles (illegal domestically) and fish species underutilized in the United States and thus has the potential for both positive and negative effects on management. Companies that service the fishing industry, providing ice or fuel, for example, are affected by the anticipated impacts of fisheries management. Small business loans to fisheries-related services are currently unavailable in anticipation of the presumed failure of much of the industry.

Management of Species and Technological Diversity

Realizing that it was impossible to manage a single groundfish species at a time, the New England Fishery Management Council developed the Multispecies Fishery Management Plan in 1986. For a time, quotas, zero bycatch tolerance, and closed areas dominated the controls on fishing, but evidence of blatant cheating, high levels of discards, misidentification of fish (e.g., cod sold as pollock to evade the quota), and lack of enforcement led to early changes in the plan. Thereafter, indirect controls on fishing were relied upon: larger mesh sizes and closed areas predominated. Stock abundance levels continued to diminish.

Some observers of the New England fisheries blame the government for overcapitalization of the industry—low-interest, government-backed loans led to too many boats and too efficient use of technology—and thus for the failure of management and the resultant downward trends in the stocks. Others blame the fishermen, citing greed, cheating, and lack of adherence to existing regulations. Weak enforcement, a government failing, is another factor blamed.

For a long time, fishermen did not believe the scientists' assessments that indicated downward trends in the stocks. There was little understanding of assessment techniques and even less comprehension of the highly technical results reported in academic journals or presented in stilted oral reports to the New England Council. After seeing the research vessels at work, fishermen would often make remarks such as, "Why don't they fish over where we know the fish are? They put their nets out where there have never been any fish and then announce the stocks are going down." Fishermen also pointed out that their landings were lower because they were using larger mesh in the nets (a mandated conservation measure), but that the scientists used the lower landings data to argue that the stocks were in worse shape.

Once fishermen began to see diminishing catches that could not be simply attributed to the larger meshes, they tended to blame pollution, global warming, fishermen using other gear, fishermen from other ethnic groups, and the hired captains of investors' boats. Now, most fishermen agree with the judgment that the industry is overcapitalized and the stocks have been overfished. They offer some support for effort limitations, but it is essential to them that any new regulations be perceived as equitable. Only equitable regulations will be supported and adhered to by fishermen.

Not all who are involved in fisheries are satisfied with the way the councils work or with the constitution of the councils. In interviews, many fishermen indicate that they do not feel well represented either by council members or by fishermen's representatives. Many feel that people who do not understand fishing, acting on advice from scientists who do not really know the condition of the stocks, are imposing management techniques that will not work and that will put hard-working, honest fishermen out of business.

The different perceptions of the New England Council's functioning and the management of the groundfish fishery have a lot to do with the relationship between local institutions and federal ones. In the next section, we briefly describe the diversity of fishermen and fishing communities in the region. Then we turn to Gloucester, Massachusetts, as a case study to illuminate the local–federal nexus. The case study presents how those involved in the groundfish fishery in Gloucester view the current crisis.

Human Diversity

According to sociologist Georg Simmel (10), the individual in the modern world is entangled in a "web of group affiliations," each making varied demands for commitment. He suggested that class, gender, religious, and regional identities are all important attributes used in determining ethnic identity. In the fisheries context in New England, an individual's identifying attributes are also tied to gear, vessel size, vessel ownership, fishing styles and skill, home port, and marketing systems.

The three preeminent groundfish fishing ports in New England have fleets of different ethnicity: "Yankee" in Portland, Maine; Portuguese in New Bedford, Massachusetts; and Italian in Gloucester, Massachusetts. In the latter two ports, many of the fishermen are American-born or have lived here for many years but maintain their ethnic affiliation or identity.

No notable corporate presence in vessel ownership occurs in New England. The majority of vessels are owner-operated or are one of two or three vessels owned by someone deeply involved in the fishing industry. Kinship relationships among crew members are common, though more so in the Portuguese and Italian communities and in rural areas than on Yankee boats operating from several urban ports. Various marketing arrangements are available, including auctions, dealers, and trucking options; urban areas tend to display more marketing options than rural areas.

Within each port along the coast, fishing intensities run the gamut from part-time income supplementation to day trips to trips lasting 10–14 days at a time. In the past, those who consistently brought in the largest catches in their category attained the prestigious title of "highliner." With the present poor conditions of the stocks, however, less positive attribution is accorded highliners. Some of the fishermen who specialize in smaller catches and emphasize high-quality handling argue that it is the highliners who damaged the fisheries and who should bear a greater share of the burden of restoring the stocks to sustainable levels.

Gloucester

Native Americans, British, Nova Scotians, Scandinavians, Irish, and Portuguese have all been part of Gloucester's maritime past. Gloucester's fishing industry is currently dominated by first- and second-generation, U.S.-born ethnic Italians. These fishermen's forebears left their small fishing villages in Italy (primarily in Sicily) when they could no longer make a decent living, when fewer fish were being caught, and when costs and regulations had increased. In a pattern woven by other immigrants in other industries, the Italian fishing families pooled their resources to buy boats and sponsor the immigration of their relatives. Those who could not work on vessels found jobs "lumping" (unloading the catches), baiting hooks, packing fish in the canneries, or working in any of the multitude of jobs that supported the industry. Thus, the extended family has been one of the most important institutions in Gloucester's fishing industry. Full employment of families in the industry may also have had the unacknowledged effect of causing the community as a whole to increase pressure on fish stocks as the stocks began to diminish, in order to keep everyone working.

Georges Bank has long been the destination of choice for captains of midsized vessels 75–90 feet (23–27 m) long. Most of the Gloucester boats that fish on Georges Bank are draggers (trawlers). Boat owners are very often the captains of one of their boats. Few individuals own more than two vessels and most are small businessmen, proud of their role in providing a healthy food product to the market.

The Italian community of Gloucester considers fishing "a way of life." Although many aspects of fishing have changed over the years, fishermen retain a sense of independence, a feeling of hardiness and bravery associated with earning a living in a harsh environment, and an emotional investment stemming from participation in an unbroken familial and cultural tradition carried on here and in Italy. Other researchers have pointed out that this strong family involvement in the fishing industry creates a resistance to movement out of the industry (11). Today, some vessels maintain four or five crew members, but one stays home each trip (rotating turns) in order to spread out the work.

The fishing tradition is fully intertwined with religious faith in Gloucester. The Blessing of the Fleet is an important part of St. Peter's Fiesta, an annual Roman Catholic celebration. For the festival, a cathedral facade is erected in the town square and an open-air mass is held. Following the mass, a parade of floats depicting various saints winds through the town. One float always depicts St. Peter and St. Paul, the fishermen who became apostles of Jesus Christ. The Catholic faith thus reinforces the fishermen's occupational choice.

Italian, or to be more precise Sicilian, dialects remain the first language for many fishermen. Formal education among those of middle age and older is often limited to the legal requirements (i.e., 5th grade for immigrants, 10th for those raised in the United States), though several fishermen are college educated. Language binds the Sicilian portion of the fishing community together, but can also function as a barrier between the fishermen and outside organizations or institutions, such as the New England Fishery Management Council where stock assessments are debated.

The women of Gloucester's fishing families have played a critical role in binding the community together. Their traditional roles have led them to organize the community in response to the fisheries crisis. Until the 1994 emergency closure of much of Georges Bank by NMFS, fishing trips averaged 10–14 days including 24 hours' (round-trip) steaming time to and from the fishing grounds. (Trips were even longer in the past when salting was done on board.) When the fishermen returned, they were often home for only a few days. Ashore, women became the heads of their households, acting as mother and father to their children. They attended Little League games and Girl Scout cook-outs; participated in mothers' and fathers' dinners and PTA meetings at school; cooked, cleaned, provided discipline and love to their children; maintained the links with relatives on both sides of the family; paid the bills; repaired the plumbing; replaced sparkplugs in the family car; and took care of the myriad other tasks involved in maintaining a home and family.

In addition to the household and familial responsibilities, some women traditionally "kept the books" for the fishing vessel, recording prices received by the boat for its harvest (the value of the catch) and the costs incurred and writing checks to the crew for their shares and to the various companies that provisioned the vessel for the trip. Today, bookkeeping is more apt to be done by settlement houses (professional bookkeepers that specialize in fishing boats), but wives may take on more of the intermediary responsibilities—picking up gear parts, arranging food, etc. In essence, they have become the "shore captains."

In Gloucester, women have traditionally maintained the networks that bound the Sicilian fishing community together. Furthermore, because they are active in the public sphere, interacting with non-Sicilians (and perhaps because they did not leave school at the earliest opportunity to go fishing), their language skills in English are often better than those of the male members of their households.

But Gloucester is by no means homogeneous. Although it is estimated that at least 40% of the community's employment and revenue depends on the fishing industry, fishermen only make up about 8% of the employed labor force of Gloucester (approximately 600 fishermen of 7,290 employed males) (12). Even within the industry, lobstermen, gillnetters, and jiggers are likely to be of ethnic origins other than Italian or Sicilian.

Ethnicity is often associated with particular institutions such as religion or language. In an urban area, ethnicity may be the group identifier that functions in the same way as "community" functions in a smaller, more homogeneous locale. However, research on ethnicity has demonstrated that individuals may select the ethnic group they identify with in order to strengthen their claims in competitive situations. Or they may assign themselves to an alternative ethnic group to create an us–them dichotomy. Such processes might be appropriately referred to as the social construction of ethnicity.

In the context of fisheries, in which stakeholders compete for scarce resources, diversity means that the stakeholders may use their differences to promote their own interests above those of "outsiders," whether the outsiders be from a different community, ethnic group, gear group, or some other category. Nevertheless, when efforts are made to reach out across the boundaries to create common goals through collaborative problem solving, such efforts may be successful.

Change in Gloucester

Today, the Italian community of Gloucester remains closely knit, but changes in the environment, physical and regulatory, have caused some unraveling. When fishing was still expanding, the extended families' full employment in various sectors of the fishing industry was advantageous. If someone was unemployed due to accident or illness, other members of the family provided aid. Full employment and economic stability now are at risk because of falling catches. Recent regulatory changes that restrict the fishery further threaten the ability of fishermen to survive economically in the short term. The current crisis has revealed the fragility of some local institutions. Recent interviews with fishing industry members indicate that family is being defined more narrowly and that support goes first to one's nuclear family, then moves out to other relatives as need and ability require and permit. Destabilization of the extended family means that some individuals have been forced to turn to the state's social welfare programs for help.

Fishing had provided a relatively good income for fishing families, for crew members as well as owners, particularly as compared with jobs available to those with equivalent (formal) education levels. But the fishermen's pride has been severely damaged by the crisis, particularly among those who have been unable to survive without turning to aid from social welfare programs. Owner-operators face the added burdens of laying off crew and, if they are unable to meet their mortgages, of losing their own homes. An important consequence of the crisis in the fishing industry has been the strengthening of the political organization of women. In a tradition borne of necessity when husbands spent long periods of time away from home, the wives of fishermen developed networks of friends and relatives to help cope with the daily round. Now that their husbands are facing strict fishing regulations and concomitant economic hardship, the women are using their networks not only as support groups, but also in such practical ways as lobbying government and religious officials for aid.

In 1969, with the help of two workers from a local community service agency called Action, the networks among Gloucester's shore captains were formalized into the Gloucester Fishermen's Wives Association (GFWA). Since then, the GFWA has pursued several routes to support the fishing industry. To encourage use of underutilized species, for example, the members developed recipes and held cooking demonstrations. When Georges Bank was being considered for oil drilling in the late 1970s, the GFWA joined forces with the Conservation Law Foundation to fight the leasing. The women lobbied (on environmental grounds) against dumping of old tires to create a reef, and they lobbied against the building of a shopping mall on prime waterfront property, but they actively supported the Massachusetts Land Bank in the development of a state fish pier.

When the members of the GFWA first became politically active, they were not accepted as appropriate spokespersons. The fishermen themselves were prejudiced against women taking public roles and, in accord with the old superstition that women were bad luck on board vessels, they did not consider the women knowledgeable enough about fishing to speak effectively for it. However, as more threats to their livelihoods arose and they realized that they did not have enough time to attend meetings themselves, many fishermen became grateful supporters of the women who could speak for them while they were at sea. Today, the GFWA tries to send a representative to all meetings of the New England Council and of industry advisory groups to speak for the interests of Gloucester fishermen.

The respect that both the government and the Gloucester community have for the GFWA became manifest when the GFWA president was hired to work for the Fishermen's Assistance Center, a multiservice human service agency established in 1994 to help fishermen and their families affected by the crisis in the fisheries. In addition, Bernard Cardinal Law of the Catholic Archdiocese of Boston has assigned a staff person to work with the GFWA on issues stemming from the downturn in the fish stocks and the stricter management regulations. Recently, GFWA became the lead agency in an attempt to draw the community of Gloucester together in order to develop a vision for the future of Gloucester's fishing industry. This collaborative problem-solving effort parallels several related efforts in the region to develop a long-range vision. The process aims to involve stakeholders from both the fishing community and the larger community in joint planning for the future of fishing in Gloucester.

Despite the ties that bind the Italian fishing community together, countervailing forces are threatening its unity. Today, gear is sophisticated and expensive, the costs of outfitting a vessel are high, and fuel and insurance costs have soared, but fish landings have plummeted. The fishermen of Gloucester have always prided themselves on their (hard) work ethic. They have followed the capitalist tradition of U.S. business, improving their technology and efficiency in hopes of maximizing their profits by improving their returns. In the context of the capitalist ethos, they benefitted their communities and the nation as they built up their businesses to the benefit of their families. However, each individual, acting in personal best interest and even in the interest of the local community, contributed to overfishing the stocks on which he or she depended. As a group, Gloucester fishermen were too successful in harvesting the species desired by the American market. They now fear loss of their livelihoods, their vessels, and their homes, many of which are collateral for vessel loans.

With the wisdom of hindsight, fishermen are being bitterly blamed in the media for overfishing. They are accused of having "killed the goose that laid the golden egg," of having been short-sighted in their greed and anxiety to reap immediate benefits to the detriment of the long-term health of their industry. Even in Gloucester, social organization provides the pattern for laying blame: fishermen themselves are divided—by gear, vessel size, species specialization, and fishing style, as well as by religion and ethnicity. Many fishermen blame the "other" fishermen for failure to conserve. Outsiders accuse the Sicilian fishermen of Gloucester of being serious offenders who are interested only in "today" because (these outsiders believe) their long-term goal is to return to Italy.

Gloucester groundfish fishermen themselves tell a different story. They speak their love of fishing despite its hardships and dangers. They talk of their children and how they have passed their commitment to the fishing industry on to the next generation. They admit that during the boom period, some people entered the fishery only to exploit it for as much as they could gain and were sometimes serious violators, putting small-mesh liners in nets, for example, or ignoring closed areas. Now, however, only the "true" fishermen, those "born to fish," are said to remain, with the possible exception of one or two "bad apples." (As will be discussed in the case study from Atlantic Canada, the ethos changed in response to scarcity.) The Gloucester fishermen talk of their distress when they find too many juvenile fish in their nets and of their efforts to convince all others to move off the juveniles. They also speak of missed opportunities, opportunities for improved fisheries management that were lost in bureaucratic requisites. They lay a lot of blame on the fishery management system. For example, Hall-Arber was told that when the Magnuson Act was enacted, some Gloucester fishermen argued that there should be a moratorium on permits to fish, that the mesh size of nets should be larger to allow more juvenile fish to escape, and that steps to assess the conditions of the stocks should be taken before the industry was allowed to expand. Gloucester's suggestions were ignored.

More recently, Gloucester fishermen lobbied to have regulators ban pair trawling, because they found it devastatingly effective, and to impose immediate closures of spawning areas (13). Eventually pair trawling was banned. (The ban was opposed by some fishermen in southern New England waters who regard pair trawling as clean and efficient. Judging from reports of the use of pair trawls for catching tuna, the technique may be very appropriate for some species in some locales [14]. Whether or not pair trawling is appropriate for groundfish has not been systematically tested.) But there have been long delays in enacting other management regulations because guidelines did not permit the incremental introduction of regulations, but instead required the whole amendment of the New England Fishery Management Council's groundfish management plan to be submitted at one time.

In general, fishermen in Gloucester do not consider the management regime fair and just. This view stems from their opinion that they are not "heard" in the management process, that fishermen's opinions are rarely considered, and—even if they are eventually considered—that the process is so slow and complicated that subsequent actions are too late to be effective. Another criticism often heard is that regulations that are imposed are not enforced. The close ties among the Italian fishermen create a reluctance to report observed violations; nevertheless, the fishermen would approve of enforcement actions against violators. Fishermen also urge that sanctions be serious enough that those tempted to violate the regulations would not do so as "part of the cost of doing business"; most suggest that permits should be forfeited for violations.

In addition, many in the fishing community feel that the management regime unnecessarily precipitated an economic and social crisis by imposing a tight schedule for recovery of the stocks. Such a schedule was mandated in the agreement that resolved a law suit brought by the Conservation Law Foundation against the U.S. Secretary of Commerce for failure to conserve groundfish stocks. Although all agree that stricter management regulations are appropriate and necessary, many believe that because the overfishing occurred over a long period, restrictions should accommodate a longer period for recovery, so that the fishing "way of life" can survive. While recognizing that the fish will rebound faster with more severe restrictions early in the process, the fishermen fear that unless the restrictions are phased in more slowly, few traditional fishermen will remain financially viable long enough to be able to fish the recovered stocks.

Homogeneity and Comanagement

Academic students of fisheries management commonly assert that it is easier to enforce regulations that fishermen support than regulations they do not support. The difficulty of enforcing regulations at sea causes managers to rely to some extent on the "honor system" and peer pressure among fishermen. Top-down management has had difficulty developing regulatory techniques that fishermen support, and recent studies of fisheries management policy have begun to look more seriously at comanagement as an alternative technique.

Evelyn Pinkerton defined comanagement as "power-sharing in the exercise of resource management between a government agency and a community or organization of stakeholders," and she described how barriers to comanagement can be overcome with alliances of stakeholders and with the help of the courts, the legislature, public boards, citizen's initiatives, and appeals to general public interest (15). A conspicuous example of successful comanagement has been the Lofoten cod fishery of Norway (16). Ostrom's studies also suggest that "self-organizing" is a potentially powerful tool in managing common property resources (17). Shared input through comanagement (or other procedures) tends to lead to rules perceived as fair and equitable by the users. When fishermen are faced with fair rules, they consider making a commitment to follow the rules so long as most similarly situated fishermen adopt the same commitment and so long as long-term benefits exceed those from a short-term, dominant strategy. However, if too many fishermen break the rules, the whole system collapses.

These studies and experiences all point out that individuals exploiting common property resources will not choose strategies that are mutually beneficial if they cannot trust each other, communicate regularly, form binding agreements, and arrange for monitoring and enforcing mechanisms. Such deterrents presently are obstacles to groundfish comanagement in New England. New England has tremendously diverse fisheries, harvested by an equally diverse population with a wide assortment of gear and fishing styles. In the past, fishermen tended to communicate best with others from their own community—those with whom they shared a trust based on similar ethnic roots, language, and attitudes towards gear and fishing styles.

Still, potential exists for defining values and goals common to all New England fishermen. When forced out of their preferred grounds, fishermen travel long distances to reach other fishing grounds. There, they find ways to communicate with fishermen from other communities who often have different attitudes, gear, and fishing styles. When danger threatens, all fishermen come together to provide aid. Ultimately, there are aspects of fishing that seem to appeal to and unite all fishermen.

The challenge, then, is to identify both the different and the common needs and goals of a heterogeneous population so that negotiations can lead to binding agreements that include monitoring and enforcement mechanisms. Various researchers have pointed out that the most successful efforts are those in which all stakeholders are involved in the negotiation and planning. Although regional fishery councils solicit the views of fishermen on management plans, fishermen have too small a role in the planning itself for this to be an example of comanagement, and they have virtually no participation in enforcement. Furthermore, the full range of participants in the industry is not represented. In contrast, the groundfish fishery of Lofoten, Norway, also has substantial diversity, but there elected representatives of gear groups from relatively small territories actively negotiate management and enforcement in a process that boasts a 90-year history of success.

The recovery of the New England groundfish fishery is further complicated because groundfish stocks range across political boundaries. Canada and the United States have quite different management regimes that affect the same stocks. Fishing communities in Canada share many features with small, rural fishing ports in the United States, but the larger institutional context changes how the crisis is viewed and the response is designed in Canada.

ATLANTIC CANADA

The Atlantic Canadian fishery is at least 100 years older than its counterpart in the northeast United States. The earliest known documentary account is a letter to the Duke of Milan dated 18 December 1497, wherein the author refers to a conversation with John Cabot who had returned to England following a voyage to Newfoundland: "the sea is covered with fishes, which are caught not only with the net but with baskets, a stone being tied to them in order that they may sink in the water. And this I heard the said Master John relate" (18). Viking and Basque whalers and fishermen may have established seasonal shore bases in Labrador and Newfoundland hundreds of years earlier still.

The patterns of human settlement and economic development in Atlantic Canada differed considerably from those in New England. In Newfoundland and Labrador, where the English achieved an early dominance, the powerful London-based merchant companies managed to have permanent settlement banned by law for nearly 200 years. The merchant companies guarded this restriction on the grounds that an indigenous fishery would infringe upon the monopolies granted them by Royal Charter. Elsewhere—and particularly in Nova Scotia—the political situation was much less settled, and both England and France encouraged settlement as an adjunct to their military struggles for the control of Canada. The fall of Quebec to British forces and the Treaty of Paris in 1763 essentially ended France's political presence in North America but left behind a significant legacy of French language and culture—most notably in the Province of Quebec, but also on the "French Shore" of southwestern Newfoundland, the Acadian French communities of northern and western Nova Scotia, and the eastern shore of New Brunswick on the Gulf of St. Lawrence.

While the New England fishery quickly concentrated in a few major ports close to the emerging centers of colonial administration, industry, and commerce, the Atlantic Canadian industry remained highly dispersed and essentially rural. The Canadian centers of political power were established far to the west in Montreal, Toronto, and Ottawa, leaving the fishing industry firmly in the arbitrary grip of English merchant companies whose interests were administrated by regional and local representatives or "factors." By comparison with New England, there was little or no industrial development to provide alternative employment for inhabitants or much of a local market for their fishery products. Until the arrival of large-scale commercial refrigeration and freezing technology in the Twentieth Century, the bulk of the Atlantic Canadian catch was salted, dried, and shipped by the merchant companies to markets in Europe and the Caribbean.

The two exceptions to this pattern were in Halifax and Lunenberg, Nova Scotia. Halifax flourished as a regional commercial and administrative center largely by virtue of commanding military control over the most northerly ice-free port on the east coast, a port that was hundreds of kilometers and many days sailing time closer to Europe than any other Canadian or U.S. port and from which a naval force could exercise effective control over the trans-Atlantic shipping lanes. Lunenberg, by contrast, had no such obvious natural advantages. It was settled in the early Eighteenth Century by farmers from Germany, some of whom turned to the sea for a living and whose skill in the design and building of fishing schooners for the Grand Banks brought them and their port lasting fame. Second only to the maple leaf flag as an internationally recognized symbol of Canada is the Lunenberg-built schooner *Bluenose*, which never lost a race with any of its challengers, Canadian or American.

Despite their historical differences and substantial differences in the philosophy and structure of their management—about which more in the following section—the fisheries of Atlantic Canada and New England share one overriding similarity: a collapse of major stocks so severe that it threatens to erase small fishing-dependent communities from the map and to shred the socioeconomic fabric of the entire region. A few simple numbers suggest the dimensions of this crisis in Canada. In 1988, the total allowable catch (TAC) for all groundfish species in Atlantic Canadian waters was 1,157,000 tons. By 1992, the dawn of the crisis, the total quotas had been nearly halved to 681,300 tons. Two years later, they were slashed to a mere 222,300 tons.

By far the most devastating effects were felt in Newfoundland and Labrador, which depended most upon the fishery and which already had the lowest per capita income and highest unemployment rate in the region. There the catches of Atlantic cod stocks fell from 245,000 tons in 1988 to zero in 1993 following the declaration of an indefinite moratorium on all cod fishing. While the available groundfish quotas fell by 80%, the numbers of fishermen and plant workers in the region stayed relatively stable at about 60,000 in each employment category (19). This persistence reflected the lack of any real or perceived employment alternatives more than it did an attachment to a "way of life."

A few islands of viability and even prosperity remain in Atlantic Canadian fisheries. Among them are the herring fishery in Nova Scotia, the American lobster fishery in general, the snow crab and northern shrimp fisheries in northern Newfoundland and the Gulf of St. Lawrence, and the offshore scallop fishery in Nova Scotia. Of these, the success of the herring fishery appears to be due to a sudden pulse of abundance after a prolonged decline, and the lobster fishery seems to be governed more by natural cycles than by human agency. Harvesters in the crab and offshore scallop fisheries have somewhat serendipitously forged excellent relations with the federal Department of Fisheries and Oceans that have resulted in de facto comanagement regimes featuring exclusionary licensing policies and very conservative harvesting strategies.

Fisheries Management

Atlantic Canadian fisheries have always been far more important in the socioeconomic and cultural life of their region than have New England fisheries, and they have been accorded a comparatively more important place in the national political agenda. Whereas the U.S. National Marine Fisheries Service (NMFS) is buried three levels down in the Department of Commerce, the Canadian Department of Fisheries and Oceans (DFO) is an autonomous administrative unit and the Minister of DFO ranks with the "Inner Cabinet" of government. Canadian management of its fisheries resources is also more centralized and absolute than it is in the United States. Federal authority in Canada begins at the waters' edge and the Minister, under the authority of the Fisheries Act, enjoys an absolute discretionary power over policy and regulation. The Atlantic provinces control the processing sector, but they have no legal authority in the management of their adjacent marine resources and their wishes in such matters are heeded or ignored at the pleasure of the Minister and the government of the day. The Minister has final authority for regulation of marine fisheries including the setting of quotas, allocation of those quotas among competing fleets and gear sectors, control of the kinds of gear and vessels employed, setting of open and closed seasons and areas, and issuance and cancellation of fishing licenses. The Minister is advised in these decisions by appointed councils and DFO staff experts but has no statutory obligation to heed their advice. In comparison with the regional council system of the United States, the Canadian system enjoys the theoretical advantage of greater speed and flexibility of decision making at the price of much less formal input from and accountability to fishery stakeholders and general citizens (20).

The same distant-water fleets from foreign nations that decimated the fish stocks of the U.S. continental shelf from the 1950s to the 1970s did similar, perhaps even greater, damage to the stocks on the Canadian side of the line. This damage prompted Canada to declare a 200-nautical-mile (370-km) management zone in 1976 and to implement it the next year contemporaneously with the U.S. action. Although Canadian management has not had to contend with the sort of resident ethnic issues and conflicts found in Gloucester and New Bedford, Massachusetts, it has been faced with other forms of ethnic conflicts, both internationally and domestically. The Canadian action on the 370-km limit was complicated by diplomatic issues with Spain and Portugal, which had been sending large fleets to the Canadian fishing banks since their discovery by Europeans and who had a strong and legitimate claim of historical dependence upon the resource. This was resolved by allocating special, annually decreasing, quotas to the two nations in return for their acknowledgment of Canadian sovereignty. A second strong challenge to Canadian management authority came from within the country as groups of Native Peoples asserted in a series of court cases that their treaties with the British Crown, and thus with Canada, had reserved for them the rights to fish as and when they pleased in their usual and accustomed places. Their claim was upheld in a landmark 1990 decision by the Supreme Court of Canada, the so-called "Sparrow decision," which ruled that Native Peoples were entitled to fish "for food and ceremonial purposes" exempt from DFO regulations except insofar as the sole intent of the regulation was conservation of the resource (21). The impact of this decision has been felt most strongly in the salmon fisheries on the Pacific coast, although it has recently become a literally combustible issue in Nova Scotia; several unregulated lobstermen of the Mi'kmaq Tribe, claiming that their entitlement includes commercial fishing, have been victims of arson.

Several important differences in the institutional structures of U.S. and Atlantic Canadian fisheries are attributable to differences in the political structures and philosophies of the two nations. By comparison with the United States, Canada is relatively socialistic and its citizenry accepts a much more centralized and active governmental role in the socioeconomic life of the country. Accordingly, the distinction between "inshore" and "offshore" segments of the fishing industry is not merely an ad hoc convention of language, but a matter of official policy, "inshore" vessels being defined as those less than 65 feet (20 m) in overall length and "offshore" vessels being longer than 100 feet (30.5 m); vessels between 20 and 30.5 m make up the "middle-distance" fleet.

Until the development of relatively small-scale "draggers" in the 1960s, inshore and offshore fishermen were also separated by gear type as well as vessel size. Inshore fishermen traditionally made day trips with fixed gear such as hook and line, baited "trawls" or longlines, gill nets, weirs, and cod traps. Additionally, the inshore fishery switched among several target species as each came into season throughout the fishing year. A failure of any one or two species to appear in a given year might be made up by an unusual abundance of others, and it seldom meant complete disaster. The offshore fishery was exclusively the domain of mobile trawlers, who made trips of up to 3 weeks and specialized in groundfish; cod was still the principal target. Further, although the locations of inshore fishing communities and the plants that bought and processed their catch had been chosen by original settlers because of sheltered harbors, nearby fishing grounds, and other factors, locations of the main ports and processing plants of the offshore fleet became largely dictated by regional development policies of the governments, the goal being to provide employment in otherwise depressed areas. The unintended consequence was the creation of "company towns" and communities that depended upon the offshore fishery more than the "inshore communities" depended on their fisheries; the latter communities, because of the more traditional seasonal round of fishing, farming, gardening, and working in the woods, had a more ecologically and economically diversified subsistence base.

Another consequential intrusion of government was the activist role of Romeo LeBlanc, the Minister of Fisheries during the crucial transition years from 1976 through 1979. Making no secret of his bias in favor of the inshore fishery, he codified the "inshore/offshore split," effectively banning the offshore vessels from the Gulf of St. Lawrence and the Scotian Shelf, and he established the "fleet separation policy" that prohibited the owners of fish plants from owning inshore vessels or licenses. The intent was to adjust the rules of the fishing game so that the people and communities dependent upon the traditional inshore fisheries did not have to try to compete for resources and markets with the vertically integrated, multinational offshore fishing corporations. Finally, while the offshore companies were issued and limited to specific annual quotas for each species, the inshore fleet was given a nonrestrictive "allocation," the amount of fish that DFO estimated could be caught by the sector in a given year. In practice, the inshore fleet was permitted to catch all the fish it could.

It was not long before some of the more aggressive inshore fishermen spied a loophole in this policy big enough to drive a small dragger through. In short order, they sold or beached their 35-foot open boats and fixed gear and ordered up 64-foot, 11-inch (19.8-m) steel-hulled draggers with powerful engines and winches and the latest in navigational and fish-finding electronics. Capable of fishing year-round and out to the 370-km limit, these vessels nonetheless were technically "inshore" vessels by virtue of being an inch shorter than the limit, but they could be fished where and when the captains pleased, unhindered by any quotas. These "cheater" boats first appeared in the ports of southwestern Nova Scotia, but more soon arrived in other major ports of that province and, to a lesser extent, of Newfoundland. This explosion in uncontrolled fishing power presented DFO with a very difficult management problem. Simply banning the boats would have bankrupted hundreds of individuals and small family corporations that had borrowed a half million dollars or more to build them. Banning them from the traditional inshore fishing ground would push them onto the offshore banks which, in addition to the considerable safety issues, would put them in direct competition with the corporate fleets that justifiably viewed those grounds as their exclusive preserve.

Accepting that the genie of technology could not be put back into the bottle, the government took the then radical step (in 1989) of ordering draggers shorter than 19.8 m fishing from Nova Scotia into a program of individual transferable quotas (ITQs). The idea was that the fleet as a whole would be issued a specific share of the quotas for the three most important groundfish species—cod, haddock, and pollock—and that each vessel within the fleet would be further allocated a share of the fleet quota based upon its past catch history. With the initial individual vessel quotas being too small for economic viability, DFO hoped that the more efficient and committed fishermen would buy out their less productive colleagues, thus matching the fleet's total catching capacity with the available resources.

Individual transferable quotas increasingly attract the attention of fisheries managers worldwide as a policy option for coping with endemic overcapacity in the fleets. They are also attracting the critical attention of many interest groups who challenge supporters' claims of enhanced resource and environmental stewardship and who raise serious questions about socioeconomic equity in the associated distribution of benefits and costs. Whatever their limitations might be, ITQ programs do foster the development of new bridging institutions between an often antagonistic fishing industry and the formal institutions of a government's management regime. Once the contentious question of implementing ITQs has been settled, both industry participants and the management agency have a strong vested interest in ensuring that the program is successful. Typically, this requires regular consultations between participants and management on a host of structural and emergent issues. This, in turn, necessitates new formal representative arrangements between management and the fishing industry. In our experience with ITQ programs, this process strongly favors the development of a de facto comanagement relationship between regulators and industry (22), which would alter management systems in Atlantic Canada.

Diversity of Fish and Technology

The Canadian fishery resource base is somewhat less complex and diverse than it is off the northeastern United States, primarily because water temperatures decrease to the north (see Chapter 1). Throughout the world's oceans, cold water is associated with high abundance of fewer species, whereas warm water promotes a higher diversity of species. The consequences of this difference for fisheries management, regional institutions, and local communities are important. In general, the lower species diversity allows Canadian fishermen and fleet sectors to target particular species more selectively than their U.S. counterparts can, but it reduces opportunities to switch species during stock declines.

Within Canada, the latitudinal effect is especially magnified in the offshore dragger fishery. The offshore fleet based in Newfoundland ports depends almost exclusively upon Atlantic cod, as do the company-owned processing plants and the communities in those ports. The offshore fleet sailing from Nova Scotia ports can fish the warmer waters on the Canadian side of Georges Bank and thus exploit a greater mix of groundfish including haddock, pollock, and various flounder species. But inshore fisheries are affected as well: they catch predominantly cod off Newfoundland and a greater diversity of groundfish and small and large pelagic species in the warmer waters of the Gulf of St. Lawrence, the Scotian Shelf, and the Bay of Fundy. A similar pattern holds for inshore invertebrate fisheries; lobster and crab fisheries have been minor and seasonal off Newfoundland and Labrador (but increasingly important since the cod closures), but much more important off Nova Scotia and Prince Edward Island. Scallop fisheries become important in the Bay of Fundy and south to Georges Bank.

Compared with the New England fishery until the arrival of the foreign distant-water fleets after World War II, the Atlantic Canadian fishery was relatively free of conflicts between inshore and offshore fleets or between fixed and mobile gears. After foreign offshore dragger fleets were banned within 370 km of shore in 1977 and inshore and offshore fleets were simultaneously distinguished, Canadian gear conflicts were minimized until the emergence of "inshore" mobile gear technology in the 1980s. Whatever else it may have accomplished, the federal decision to place the inshore dragger fishery under ITQ restraints has done nothing to resolve the divisive and emotionally charged confrontations between mobile-gear and fixed-gear sectors of the inshore fishery. These disagreements simmer and occasionally boil over at meetings between industry and DFO (mostly in Nova Scotia and New Brunswick; to a lesser extent in Newfoundland). The resulting conflicts have received extensive and ongoing coverage in the local and regional media.

Fixed-gear fishermen accuse the draggers of the "rape" of the resources and the destruction of sensitive benthic habitat. Mobile-gear fishermen counter that they have taken a leadership position in developing sustainable harvesting practices and other important conservation measures, including their ITQ program. The ITQ program, they explain, is a self-funded buyout of excessive harvesting capacity and includes a 100% dockside monitoring program, also paid for by sector members from the proceeds of each vessel's landings. But, reply the fixed-gear fishermen, that simply means that the draggers have an extra incentive to discard their bycatch and "high-grade" their quota species at sea. Not so, say the draggermen—we have voluntarily introduced a regulation that requires us to fish with mesh of a type and size that is highly species-selective and that reduces the catch of small fish to almost nil. In return, draggermen accuse fixed-gear fishermen of using "trout hooks" on their longlines to catch large number of juvenile cod and haddock; of using increasing numbers of small-mesh gill nets, many of which are lost through poor fishing practices but continue to kill fish (the "ghost-fishing" problem); of landing catches generally unobserved and unmonitored; and of regularly overrunning their sectoral quotas. And on it goes.

In Newfoundland, the conflict between mobile and fixed gears is not within the inshore sector, but between the primarily fixed-gear inshore fishery and the exclusively mobile-gear offshore fishery. Inshore fishermen blame offshore draggers for the collapse of the cod stocks because the draggers concentrated their effort on the dense spawning aggregations that form in the late winter and spring of each year. An organization representing Newfoundland inshore fishermen brought suit in federal court for an injunction, which was not granted, to prevent fishing on the spawning grounds. The Newfoundland offshore mobile gear sector says that the real problem is the traditional inshore cod trap fishery, which catches primarily small fish before they have had a chance to spawn. Maybe that is true now, reply the inshore fishers, but we have little left to catch after the draggers have destroyed the adult spawning stocks. And on it goes.

Meanwhile, the few hundred individuals who held licenses for snow crab tried desperately to counter the pressure on DFO to issue more crab licenses to inshore fishermen whose fisheries had closed. Depleted to the point of collapse in the early 1980s, the snow crab resource has rebounded strongly in the 1990s. Combined with the collapse of competing crab fisheries elsewhere in the world, the rebound has made the current license-holders very wealthy—so much so that in 1994 the 70 or so individuals who held licenses for the central Gulf of St. Lawrence (the most productive grounds) could donate $3 million for the relief of their neighbors holding only groundfish licenses. The crab fishery has expanded, however.

The Gulf of St. Lawrence snow crab fishery exemplifies the evolution of unintended but effective comanagement between DFO and a community. When DFO began to take notice of the developing snow crab fishery in the early 1980s, it happened that the biologist designated to lead the assessment team was Japanese and that the primary market for crab was in Japan. When the biologist began asking questions, fishermen assumed he represented a Japanese buyer and accorded him an automatic respect that would not have been forthcoming if they had known he was from DFO. By the time the misunderstanding was cleared up, the biologist had established himself in the fishermen's minds as a knowledgeable and reasonable person. Furthermore, when DFO undertook the first comprehensive assessment survey of the snow crab stock, it chartered several commercial crabbers under the command of their usual skippers. This placed DFO biologists on the crabbers' turf and terms, an unusual reversal of the normal structure of DFO–industry interactions. Excellent and cooperative relations between DFO and crabbers resulted, providing a model for improved institutional relationships in the fishing industry.

Nova Scotia Case Study

A case study of one community in southwestern Nova Scotia highlights the differences between U.S. and Canadian management policies and practices for Atlantic groundfish. The study also illustrates how local institutions and norms of behavior influence the judgments people make about the fairness and equity of management and the views they have on compliance with fishery rules. Finally, it offers some insight into the effect quasi-privatization of a fishery can have on community solidarity and levels of compliance (23).

The Scotia-Fundy groundfish fishery covers a region along the coast of Nova Scotia from Cabot Strait off Cape Breton to the U.S. line in the south. This region includes the rich fishing grounds of the Canadian side of Georges Bank, Browns Bank (northeast of Georges Bank), and the Bay of Fundy. Historically, this fishery has depended on cod, haddock, pollock, and flounders. The complex Scotia-Fundy fishery is managed by gear type and by vessel size; for vessels in the 14–20-m length range—the "middle-distance draggers" (initially called "cheaters")—recent management has been by individual transferable quotas (ITQs). This ITQ program was implemented in 1991 in response to rapidly increasing capacity and declining stocks during the late 1980s. Middle-distance draggers, the main focus of our discussion, have been allocated approximately 25% of the quotas for cod, haddock, and pollock. The vessels also target other species, including flounders, dogfish, and skates. Since 1991, the quotas have been made fully transferable, Georges Bank has been included in the management area, and ITQs for flounders have been implemented for some management areas.

The community chosen for the case study is an Acadian village on the southwestern tip of Nova Scotia. Most of the community's 2,000 residents earn their living as fishermen, as plant workers or owners, or in fishery support businesses such as boat building, welding, and fuel supply. Many are also involved in the fishery management process as members of industry advisory groups. The ancestors of many residents were among the first French settlers in the region and pioneers of commercial fishing in the area.

Before Canada and the United States extended their seaward jurisdictions, fishermen from this village regularly fished off the U.S. coast from Boston south as far as Cape May, New Jersey, and U.S. fishermen exploited Nova Scotian waters. Fish from this village are taken daily to Boston. The fish dealers in the village and New England compete in the world market for groundfish, scallops, and lobsters. Although the village shares much with fishing-based communities in New England, it is much more homogenous and much more dependent on commercial fishing than they.

Within living memory, several Scotia-Fundy fisheries have boomed and busted, beginning with scalloping in the 1960s and followed by herring seining, groundfish fishing, and lobstering. Most dramatic was the boom in the herring seine fishery in the late 1960s and early 1970s. Herring vessel skippers made more than $100,000 a year and deck hands half that. The younger men, particularly, then used the experience and money gained to invest in the middle-distance dragger fishery. A few draggers worked out of this port in the 1960s and early 1970s; by the mid-1970s, some fishermen had installed more powerful engines, modified trawl doors and nets, and built bigger vessels. As the word spread that dragging could be more efficient than longlining, more fishermen converted longliners to draggers. By the early 1980s, experienced owner-operators were building state-of-the-art draggers.

Skippers of this generation also were more aggressive than their fathers, working 7 days a week, winter and summer, regardless of the weather. The introduction of loran C navigation allowed them to drag their nets in spots their fathers had had to avoid for fear of ripping their nets on rough bottom. Information about these locations always spread quickly through the fleet, thanks to the tradition of "yarning" (shore-side discussions of each day's trip with other captains and hands) and the unspoken rules that required sharing of information among friends.

After 1977, and particularly in the 1980s, the Canadian government encouraged growth of fleet capacity through quota management and generous fishery development loans. These government programs created a kind of feedback loop in which the fishermen constantly out-distanced managers' attempts to control fishing effort. Many landings went unreported in this village and throughout southwestern Atlantic Canada; fishermen estimate that villagers were landing 25–50% more fish than were being reported. Because they were all doing it, fishermen did not consider cheating to be "breaking their own rules." Rather, they were successfully adapting to outsiders' (fishery management) rules by circumventing them. The government's attempts at controlling effort, such as trip limits, merely changed the rules by which the fishermen won the game.

Rapidly improving technology in fish-finding equipment, larger and more powerful vessels, and increased fishing effort helped to mask the declines in stock levels. Money continued to flow from federal and provincial agencies, as well as from banks, for boat building and improvements to fish plants and for buying or upgrading vessels. A generous unemployment system made it possible for fishermen to work seasonally and to live on unemployment insurance during the rest of the year. Those who did not go fishing could qualify for unemployment insurance by working in the fish plants. The combination of an apparently unlimited fishery resource, access to world markets, generous government support programs, and ingenuity and hard work made the village prosperous.

As stocks declined through the 1980s, fish prices rose, and more vessels entered the fishery. The government imposed various controls, such as trip limits, to hold the fleet to the quota levels, but the controls did not prevent overfishing. The government finally closed the fishery in June 1989. With the closure, the fishermen realized that the balance of power—which had always been with the government on paper—had moved toward the government in practice as well.

Canadian government proposals for reducing fleet capacity after the 1989 closure were based on a new study showing that the Scotia-Fundy fishing industry was threatened by declining stocks and increasing fishing and processing capacity despite the reductions in total allowable catch that had been made (24). In December 1989, the Minister of Fisheries and Oceans announced that ITQs would be used to reduce capacity in the middle-distance dragger fishery. The government then appointed an Industry Working Group to recommend the formula for initial ITQ allocations, the appeals criteria and process, the extent to which ITQs would be transferable, and the amount of quota to be assigned to each license holder. The recommendations were put on a ballot sent to license holders.

In January 1991, an ITQ system based on historical catch was implemented for cod, haddock, and pollock in the southwestern Nova Scotia region. The system required other management measures as well, notably a dockside monitoring program with 100% coverage, mandatory vessel logbooks, onboard observers, and heavy administrative sanctions. In addition, no license holder could acquire more than 2% of the total cod, haddock, and pollock quota. Fishermen fishing with draggers under 14 m could choose to join the ITQ system or to pool their quotas into a "generalist" category. Initially, the ITQ system was optional for mobile-gear fishermen in eastern Nova Scotia, who could instead opt to fish on a competitive quota. Since implementation, the ITQ system has expanded to include Georges Bank cod, haddock, pollock, and flounders. Groundfish species not included in the ITQ system are managed by a competitive quota, but are subject to the comprehensive dockside monitoring program.

Village Adaptation to ITQs

Fishermen from the village attended the government's industry consultation meetings on ITQs. Early meetings were acrimonious. Most fishermen from the village violently opposed any sort of ITQ system on principle: it would shut out young fishermen and favor those with lots of money, and there was no way to design the initial quotas that would be fair to everyone. However, when it became clear that the government was determined to impose an ITQ system, fishermen in the village became determined to adapt successfully to it. These fishermen tried to learn all they could about how these regimes work, and began acquiring vessels that were likely to benefit from the ITQ system.

Those who had the money or could borrow it bought quotas from others once the regime was implemented (initial quotas were not expected to be economically viable), starting a process that has concentrated ITQs in the hands of a few companies with close connections to the fish plants. Currently, only 2 or 3 of 30 vessels in the middle-distance fleet remain independent of the fish plants, whereas approximately 20 independent operations existed before 1990. Quota holders have acquired substantial power in the community. Not only do they determine who among the pool of crewmen can go fishing, they also determine how much fish the crew can catch. Some of them also have bought other limited-entry licenses, such as those for shrimp, particularly if they had more than one boat and not enough groundfish quota to keep them fishing. Quota holders have also landed "underutilized" species such as redfish and yellowtail flounder if they could find a market for them.

Quota owners believe their investments in ITQs bring significant benefits to the fishery and to this community by financing an industry buyout of excess fishing capacity and by buying ITQs from outside the village so that the fish are landed in the village rather than elsewhere. Nevertheless, the concentration of fishing rights in this egalitarian and close-knit village has brought severe social stresses. Quota holders have sought to mask the increasing social and economic stratification by continuing to "yarn" in coffee houses and other gathering places. Non-ITQ holders have sought to maintain the social status quo by engaging in traditionally fierce debate with everyone, quota holder or not. When ITQ holders seem to be exercising too much power, they are criticized directly, as is the ITQ system itself.

A common argument is that the ITQ system rewards those who own the fishing rights, rather than those who catch the fish. Quota owners have been labeled "fish lords" and "armchair fishermen."

But rules of politeness and norms of equity cannot overcome economic differences entirely. A few ITQ holders have become decidedly more assertive about their ownership of the fish caught, and some crewmen have been threatened with firing if they do not perform strongly. The practical exercise of economic authority limits the effectiveness of social efforts to maintain status quo.

Fishery Value of ITQs

When discussion of the ITQ system turned from its fairness to its success in preventing overfishing, villagers indicated that the new regime had reduced overfishing by making quota management more effective. Any ITQ regime directly allocates to each ITQ holder specified amounts of fish. The Scotia-Fundy regime requires 100% dockside monitoring of all vessels with ITQ licenses, administrative sanctions (license revocation or loss of the ITQ allocation for that year), and a square-mesh net for cod, haddock, and halibut (a requirement sought by the industry under the belief that square-mesh netting allows more undersize roundfish to escape than diamond-mesh netting).

A 1993 survey of inshore Scotia-Fundy draggers conducted by DFO indicated that respondents were about evenly divided on whether illegal fishing had become worse, better, or had stayed the same under the ITQ system (25). People in the village, including crewmen who did not own ITQs, tended to be more optimistic. They said that dockside monitoring and sanctions were preventing most types of cheating. Even discarding at sea was down, which was attributed to lower retention of small fish by the square-mesh nets.

Village Overview of ITQs

Ownership of ITQs has concentrated fishing rights into the hands of the relatively few who have had access to capital—the plant owners—in the study village. The nature of ITQs as quasi-private property has changed social relations in the village, and a trend toward social stratification is evident. The villagers, especially those who are directly involved in the fishery, consider this a significant social cost of the new regime.

On the other hand, ITQs may bring a social benefit through enhanced resource sustainability. Deckhands and captains, as well as ITQ owners, said they have become convinced that their own best interest is to husband the resource. However, they also credited the 100% dockside monitoring and administrative sanctions—not the ITQs themselves—as the most important causes of reductions in discards. They also were convinced that tough administrative sanctions and requirements for square-mesh nets, adjusted to the size that allows undersize fish to escape, have reduced mortality of juvenile fish.

The outcomes of the Scotia-Fundy ITQ system to date indicate that the fishery management process in Canada, while more autocratic than the U.S. process, holds promise for creating stakeholders in the system and for educating these stakeholders about how to serve their own best interest by promoting a sustainable fishery. At the same time, some of the fishermen interviewed pointed out that the conservation ethos highlighted by the ITQ system was reinforced by a context of scarcity and the imposition of strong enforcement measures, such as 100% dockside monitoring of catches.

The ITQ holders claim that it is impossible to manage a fishery until the stakeholders in that fishery have been specified, as they are in an ITQ system. Their honesty about their former practices and their near-amazement at the change in their own perception of what constitutes a successful fishing strategy support a belief that effective management entails the specification of those who have fishing privileges and therefore a stake in the successful management of the fishery.

SUMMARY

Despite overt differences between the city of Gloucester and the featured Scotia-Fundy village, there are some striking similarities in the institutions associated with the fishing sector of each community. Both have social organizations that place a high value on extended families and community. Both locales evidence socioeconomic stratification that arose with competition and differential financial success, despite strong egalitarian values. Both locales share a history of a fishery initially overexploited by distant-water fleets, then overdeveloped by domestic fleets with the impetus of various government grant and aid programs.

Canadians criticize U.S. fishermen for continuing to overfish despite assessments that indicate stock declines. Ironically, U.S. fishermen long were skeptical about low U.S. stock assessments because very positive assessments were reported for Canadian waters. The U.S. fishermen pointed out at the time that fish do not stop swimming at the national boundary, so they suspected that the negative U.S. assessments were inaccurate.

When management changes were proposed that would limit fishermen's access or rights to fish, fishermen in both locales voiced strong opposition. Perhaps the major difference is that U.S. fishermen continued much longer (and still do) to openly battle regulations, using their structural role in the U.S. regional fishery management council system to do so. In contrast, Nova Scotia fishermen concentrated on adapting to the ITQ system their government had the authority to impose upon them. We believe that the differences in the institutions that specify national fishery management strategies have influenced the social responses to fishery regulations.

Canadian fishermen freely admit that there was a tremendous amount of cheating when the groundfish fishery was under earlier forms of quota management, though they say that it has diminished under the ITQ system. They add that it is the entire management program, especially the severe sanctions, that has changed their behavior. Gloucester fishermen perhaps less freely admit awareness of violations in U.S. waters. They also claim that only "true" fishermen are left (i.e., fishermen who care about the future of the fisheries) so now there is very little cheating. In their view, ITQs alone do not make fishermen take individual responsibility—commitment to family, community, and the long-term health of the industry must also obtain. Further, it is said that the fishermen who were "in it to make a killing" have been scared out of the fishery by the poor stock assessments and stricter regulations. Stricter enforcement, coupled with speedy application of sanctions, is viewed as essential.

"I'll do anything I have to do to take care of my family, to keep my boat and home!" More than one U.S. fisherman said this during interviews and discussions. Although we did not examine this issue directly in the Canadian research, we sense that the national policy on employment insurance affects the willingness to cheat among fishermen in each country. In Canada, the unemployment insurance system covers men and women who fish or work in the fish plants for 12 weeks. Should the fishery not provide this minimum, it is likely that provincial make-work programs would provide enough work to meet the federal criterion. One Nova Scotian fishery enforcement officer said that this was only fair, because both the federal and provincial governments control fisheries effort. In this man's mind, and perhaps in others, government's authority to make decisions unilaterally goes hand in hand with government's duty to take care of the people affected. Moreover, we observed that Nova Scotian fishermen's self-identity does not depend on their being full-time fishermen. In Atlantic Canada, there is no stigma attached to fishermen who qualify for unemployment; they still consider themselves fishermen first, even if the majority of the year they are on unemployment insurance.

In contrast, U.S. fishermen whose incomes are threatened by government measures to restore the fish are not at all sure they will be able to bring in enough money to support their families or establish other businesses. Many U.S. fishermen fish as independent contractors and thus have no unemployment insurance, yet they are ashamed to seek welfare benefits such as food stamps or Aid to Families with Dependent Children. Currently, the social safety net in Canada protects fishing families from losing their homes and all income; the same is not true for fishing families in the United States. In addition, a lack of social stigma associated with state aid allows unemployed Canadian fishermen to maintain their sense of pride.

The larger institutional context of fisheries management differs between the two countries. In Canada, authority is without question top-down; that is, it rests with the central government. The Canadian government has been very enthusiastic about creating quasi-private property rights to limit fishing effort. If the case of the Scotia-Fundy middle-distance dragger ITQ fishery is typical, specifying stakeholders creates the institutional context for a shift away from a consultative process and towards a comanagement process. However, power and authority remain with the federal government.

In the United States, the regional fishery management councils were designed with the idea that only those involved in the industry can understand it and, therefore, they must be engaged in its management. Though the Secretary of Commerce has ultimate responsibility for accepting or rejecting management plans, the councils write the plans, take them to public hearing, and forward them to the Department of Commerce. Although the U.S. system leans towards comanagement, in practice those who have a voice are most likely to be those with the most power and money, tempered perhaps by groups who have organized themselves.

Gear restrictions, effective monitoring, and severe sanctions seem to be key requisites to a successful management plan, but Canadians maintain that stakeholders must be specified if comanagement is to be effective. The jury is out on whether or not community development quotas (CDQs) have the institutional capacity to foster sustainable resource practices. If they turn out to be effective conservation mechanisms, CDQs, or hybrids of ITQs and CDQs, may offer a means of specifying stakeholders in ways that are sensitive to community views of what constitutes a fair and equitable distribution of the resource.

In sum, community-level institutions influence social responses to federal management strategies. The Canadian study argues that local institutions inform people's assessments of the fairness and equity of the fishery management system, shape views on the morality of cheating, and help people adjust to social impacts of a management regime. Local institutions do not operate in a vacuum, however, as these studies demonstrate. Federal fishery management plans can be designed with incentives or disincentives to cheat, and they can or may not cohere with local notions of fairness and equity. Both the Canadian study and the U.S. study indicate that the public choice process also shapes local views and actions in response to fishery management. When fishermen participate in the fishery management process, they become invested in the success of the management, whether the industry's role is advisory or comanagerial.

NOTES

1. Women who fish northwest Atlantic waters generally prefer to be called "fishermen," and we use this term through the chapter.
2. Ostrom (1990).
3. McCay and Acheson (1987).
4. Pinkerton (1989); McGoodwin (1990); Ostrom (1990).
5. Smith and Bacek (1992).
6. Hennemuth and Rockwell (1987).
7. Bourne (1987).
8. Tub trawl fishermen were essentially longliners who coiled their hooks and lines in tubs or half-casks ready to be set. Anderson (1987) has described the procedure.
9. This and other historical information is contained in an unpublished paper by S. A. Murawski entitled "A Brief History of the Groundfishery in New England, 1880–1930," which was presented to the Conservation Law Foundation of Boston.
10. Simmel (1955).
11. Doeringer et al. (1986).
12. G. Atkins (city of Gloucester economist), unpublished "Fact Sheet on the Gloucester Fishing Industry, 1994; A. San Filippo (President, Gloucester Fishermen's Wives Association), personal communication, 1995.
13. In pair trawling, two boats each pull one end of a long net.
14. Goudey (1995).
15. Pinkerton's (1989) work on common property resource management was conducted in the Pacific Northwest.
16. The Lofoton fishery, along northwestern Norway north of the Arctic circle, has been described by Jentoft and Kristoffersen (1989) as "the largest cod fishery in the world in terms of catches and participation," even though the vessels and gear used are diverse and the fishing area is spatially limited. Like New Englanders, Lofoton fishermen con-

tend that "practical and local knowledge of the fishery is crucial" if the regulatory system is to adapt appropriately to variations in the fishery. In the Lofoton system, each of several defined ocean territories or districts is administered by a superintendent and staff. The skipper of each registered vessel votes for fisherman inspectors to serve on the regulatory committees, which set limits on where and when certain gear can be used. Gear groups elect their own inspectors who oversee adherence to regulations and who also elect representatives to the regulatory committees. Every fisherman must follow the regulations, and those who have participated in the decision making have more positive attitudes toward the regulations than nonparticipants. Decentralization of regulatory functions has allowed more management flexibility, which means the regulations are more apt to be perceived as appropriate and fair.

17. Ostrom (1990) explored both the theoretical literature and a wide assortment of case studies to determine ways of analyzing and resolving the "tragedy of the commons." The case studies involved relatively small-scale situations in which users of a common wanted to solve mutual problems in order to enhance their own long-term productivity. The solutions that worked in these circumstances are more difficult to apply as the size, number, and heterogeneity of user groups increases.
18. Quoted by Pearson (1972).
19. TFIAAF (1993), the "Cashin Report."
20. Apostle et al. (in press).
21. Parsons (1993).
22. McCay et al. (1995).
23. Regional management histories and details of management plans have been written by Haché (1989), Annand (1992), Angel et al. (1994), Burke et al. (1994), McCay et al. (1995), and Apostle et al. (in press). The community analysis supporting this case study was conducted by C. Creed during a postdoctoral fellowship at Dalhousie University. Interviews of community members, especially in relation to individual transferrable quotas, were summarized in an unpublished paper by C. Creed, R. Apostle, and B. J. McCay entitled "ITQs from a Community Perspective: the Case of the Scotia-Fundy Groundfish Fishery," which was read at the annual meeting of the American Fisheries Society in Halifax, Nova Scotia, during August 1994. The paper is cited with the permission of Dr. Creed, who also provided additional background about the study village in early consultations over this chapter; a copy of the paper may be obtained from Dr. Bonnie J. McCay, Departments of Anthropology and Human Ecology, Rutgers University, New Brunswick, New Jersey 08903, USA. Aspects of the community study also were reported by Creed (1996) and McCay et al. (in press).
24. Haché (1989).
25. Creed (1996).

PART II

MANAGEMENT CONSIDERATIONS FOR REHABILITATED STOCKS

Part II shifts the book's focus from the present to the future. Depleted groundfish stocks will require several years of slight or no exploitation before they recover, though no one really knows how long recovery will take, how recovery will be affected by environmental variables, or how similar the recovered stocks and fish assemblages will be to those of the fishery's heyday. In the meantime, previous management systems will be dissected and analyzed, and new management systems will be debated.

The following chapters were written to foster discussion about groundfish management among all stakeholders in the fishery. They outline a variety of future options ranging from "no change" through sophisticated population management, ecologically oriented management, gear management, and grassroots management to essentially private management. The options are not mutually exclusive; several could be pursued in tandem. The final chapter describes a decision-making system by which any management option could be planned, implemented, and evaluated.

The authors in Part II were encouraged to be "advocates" for the options they propose so that potential merits of each approach would be stated clearly and forcefully. No approach is perfect, however, and the liabilities of each have not been ignored. Whatever their approach, the authors all share the goal of a healthy groundfish resource that will support sustainable harvests into the distant future.

Brian J. Rothschild, Alexei F. Sharov, and Marjorie Lambert

SINGLE-SPECIES AND MULTISPECIES MANAGEMENT

In the early 1880s, the famous British scientist Thomas Huxley proclaimed that the great sea fisheries were inexhaustible and that attempts to regulate them were useless. Within only a few decades of this optimistic statement, however, several major fish stocks were clearly declining, evidently under the pressure of fishing (1).

The declines in fish stocks pointed to the need for fisheries regulation and maintenance of stock sustainability—in particular, for balancing fishing with conservation. If a stock is not fished, it makes no economic contribution. If it is fished too intensively, its value declines considerably. Between these extremes, fishing intensity may be regulated (in theory) to produce a biological or economic optimum yield. Attaining long-term optimal levels of fishing intensity is the principal goal of fisheries management.

Current marine fisheries management in North America has been built on biological rather than on economic criteria. Most of it has been directed at individual species or stocks rather than at complexes of co-occurring species. This "single-species" tradition arose partly because most fisheries target particular species and partly because estimating optimum yield for more than one species at once is scientifically daunting. More than a decade of work in Europe suggests that multispecies management continues to be a difficult technical undertaking (2). In the sections that follow, we briefly outline the principles, procedures, problems, and prospects of single- and multispecies management with particular reference to northwest Atlantic groundfish.

SINGLE-SPECIES MANAGEMENT

Single-species management typically is applied to individual stocks of a widespread species such as Atlantic cod or haddock (3). Its goal is to specify optimal levels of size-specific fishing mortality for a particular species. "Fishing mortality," designated F by tradition, is the instantaneous rate of death caused by fishing, and it is one of the few aspects of exploited stocks that fisheries managers can control. "Size-specific fishing mortality" is the fishing-related death rate of each size-group in the stock. Thus, management attempts to determine the level of fishing mortality (F) that will allow the optimum yield, and then to regulate the fisheries—by designating net size, mesh size, vessel power, number of vessels, and fishing seasons—to achieve that F.

Single-species management relies heavily on mathematical models that require current biological data such as fish age, length, weight, and maturity. A touchstone of all models is yield, the weight of fish landed from a stock each year. Yield is usually calculated from market and vessel records for "directed" fisheries—fisheries in which one stock is "targeted." To be most useful under the single species management concept, yield of a targeted stock should be supplemented with records of bycatch (the incidental catch of the same species in other fisheries) and discards (fish thrown overboard, usually dead or dying, because they are unwanted or of an illegal species or size). The use of gears, such as trawls, that catch many species and sizes of fish accentuates the bycatch–discard problem, which is a major one for single-species management. Bycatch and discards can account for major portions of the fish removed from a stock, but they are notoriously difficult to estimate accurately.

Most models in fishery management make use of fishing effort or catch per unit effort, "unit effort" being an hour of trawling, a 24-hour set of gill nets, or some analogous standard measure. Such annual indexes of fish abundance are obtained by monitoring commercial catches at sea or by independent research fishing. By sampling fish in commercial or research catches, biologists can estimate the size (and age) distribution of fish in the stock and the growth rates of the fish. From these data, rates of mortality and recruitment and size-specific rates of biomass production can be calculated. Recruitment refers to the number of young fish that become newly vulnerable to the fishery during the year; it is one of the most difficult values to estimate accurately. Biomass production is the weight gained (on average) by recruits before they are caught or otherwise die. Mortality, measured as the rate at which year-classes of fish disappear from a stock over time, is the total of fishing mortality and "natural" mortality (all other mortality unrelated to fishing); a fixed value is generally assumed for natural mortality, and fishing mortality is considered to be the difference between the total mortality and natural mortality.

The models variously manipulate recent estimates of recruitment, biomass production, mortality, fishing effort, and yield to project total abundance or biomass of the stock, the size or age distribution of fish within the stock, and the expected yield at the end of the next fishing year. With this information, managers can set an appropriate target level of fishing mortality F for the following year and prescribe the total allowable catch (TAC), quotas, gear regulations, and season and area closures to achieve that F. There is no technical limit on the target F that can be chosen, but the usual practice is to select one of only a few standardized mortality targets, as discussed below. These standard F values, as well as criteria for stock size, fishing effort, and other features thought to optimize a fishery, are often called biological reference points.

Three classes of models are used to set optimum levels of size-specific fishing mortality; these are called surplus production models, dynamic pool models, and stock–recruitment models. Fully calibrating these models for a stock requires data over a wide range of stock conditions from newly fished to overfished. Such data are available for most stocks of northwest Atlantic groundfish.

Surplus Production Models

Surplus production models relate yield to fishing effort (4). In a new "ideal" fishery, increases in fishing effort (more nets, more boats, more days) bring higher yields at first, although each additional net, boat, or day brings a smaller increase in yield than the previous addition. After a maximum yield is reached, determined by the species' productivity, further increases in effort bring decreases in yield, and the result is a dome-shaped yield–effort curve. Fishing mortality is assumed to be proportional to fishing effort, so the yield–mortality relationship also is dome-shaped in this model (Figure 6.1). An assumption of this fishery model is that biological and environmental conditions that might affect populations do not change and that stock size and yield will be constant if fishing effort (hence mortality) is kept constant. The model specifies the fishing mortality that would allow maximum sustainable yield (MSY); this mortality is called F_{MSY}.

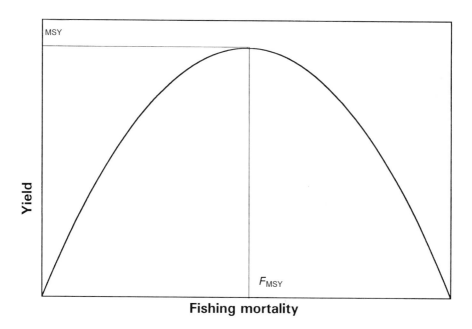

FIGURE 6.1.—General relationship between fishing yield and fishing effort or fishing mortality in surplus production models. Maximum sustainable yield under stable environmental and population conditions is denoted MSY. The fishing mortality that keeps a fishery at MSY is called F_{MSY}.

Dynamic Pool Models

Dynamic pool models generate a quantity called "yield per recruit" and they determine the fishing mortality that will maximize this quantity (5). Yield per recruit is the theoretical average harvest weight yielded by each recruit during its remaining lifetime in the fishery, and it is typically related to fishing mortality by a domed curve (Figure 6.2). Yield per recruit is obviously zero when there is no fishing, and it is low when fishing intensity is so great that recruits are caught as soon as they enter the fishery. At intermediate levels of F, some fish are able to grow for one or more years before they are caught, which increases the average yield per fish. The fishing mortality that allows the greatest yield per recruit is called F_{MAX}; fishing at a rate higher than F_{MAX} is termed "growth overfishing" (6).

Dynamic pool models make use of more information on size-specific growth and mortality than surplus production models. However, variability in recruitment, growth, and mortality is implicit in surplus production models. A practical consequence is that F_{MAX} calculated from dynamic pools may exceed F_{MSY} calcu-

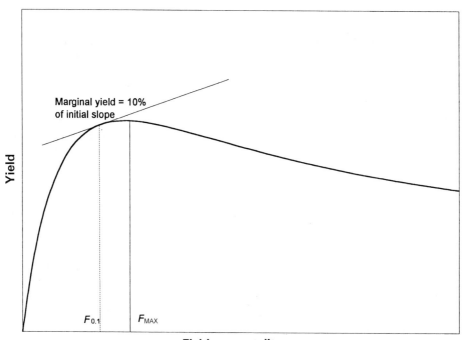

FIGURE 6.2.—Relationship between yield per recruit and fishing mortality in dynamic pool models. The fishing mortality that keeps a fishery at maximum yield per recruit is called F_{MAX}. The more conservative $F_{0.1}$ corresponds to the point at which the slope of the curve is one-tenth the slope for a newly exploited fishery. The slope of the curve at a point is the slope of the tangent at that point. The slope is greatest for a virgin fishery. It becomes smaller as the fishery intensifies (i.e., yield per recruit increases at a smaller rate) but remains positive until it reaches zero (the tangent is horizontal) at the curve's apex. Thereafter the slope is negative, meaning that further increases in fishing mortality bring declines in yield per recruit.

lated from surplus production (7). Thus, relying on F_{MAX} may expose a stock to greater risk of overexploitation than relying on F_{MSY}. In recognition of this problem, but also in recognition that the complexity and variability of real fisheries induce considerable uncertainty in model estimates, many analysts recommend that fishing mortality be set at the precautionary level of $F_{0.1}$. This is the mortality at the point on the yield-per-recruit curve where the slope of the curve is one-tenth the slope for a nearly virgin fishery (Figure 6.2). Typically, $F_{0.1}$ is slightly lower than F_{MAX} (8).

But F_{MAX} and $F_{0.1}$ give only a partial picture of optimality. The values incorporate yield per recruit but they do not account for variations in total recruitment, yet recruitment is often the most important consideration in fisheries management. The history of many fisheries has demonstrated that increasing exploitation is often associated with reduction of recruitment and sometimes a collapse of the stock (termed "recruitment overfishing"). The analysis of recruitment variability—the stock–recruitment relationship—has become one of the most important (and the most difficult) issues of fisheries management.

Stock–Recruitment Models

A stock–recruitment relationship is usually represented as a graph of stock size (total stock biomass, spawning stock biomass, number of spawners, number of eggs produced, etc.) versus recruitment produced by the stock. Two basic curves describe this relationship—a dome-shaped and an asymptotic curve (Figure 6.3). These curves represent simple ideas of how recruitment depends on stock size. Both models imply that a very small population produces small recruitment. As stock size increases, recruitment should increase because more spawners are available to produce eggs. Such an increase cannot be sustained indefinitely, because environmental limitations on food, spawning area, and other important life history variables limit the size of the stock. If the effectiveness of reproduction decreases after some maximum (for example, if overcrowded females reduce their egg production or superabundant young deplete their food supply and starve), a dome-shaped curve results. Sometimes, however, recruitment becomes nearly constant (asymptotic) after some point, no matter how abundant the stock becomes, because the environment can sustain only so many young and no more. In reality, actual data often diverge from a mathematical model and appear as a cloud of points surrounding the theoretical curve. Therefore, the theoretical stock–recruitment relationship only predicts "average" recruitment. However, with generous allowances for variability in the data, it does allow one to predict good and poor year-classes (9). This is especially important when a stock is at low level. A stock–recruitment relationship could be used to determine the minimum spawning stock biomass (total weight of spawners in the stock) that would protect the stock from collapse. The usual way of doing this is to "invert" the stock–recruitment relationship so that instead of predicting future recruitment from current stock size, we predict future spawning stock size from current recruits. The statistic used is "spawning stock biomass per recruit": the future lifetime weight of spawner produced by each current recruit.

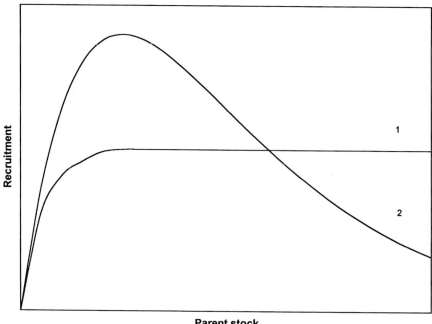

FIGURE 6.3.—Stock–recruitment curves based on **(1)** Beverton and Holt (1957) and **(2)** Ricker (1954). As parent stock size increases, curve (1) approaches but never quite reaches a maximum level of recruitment (the asymptote).

Spawning Stock Biomass per Recruit

Spawning stock biomass per recruit (SSB/R) is an estimate of the average weight each recruit will reach after it becomes sexually mature but before it dies (in the fishery or otherwise). This biomass is affected by growth rate (the faster the growth, the greater the biomass), natural mortality (the higher the mortality, the smaller the average weight achieved before death), and fishing mortality (the greater the exploitation, the smaller the average weight). A larger average size (weight) has several positive implications for reproduction within the stock; larger fish produce more eggs than smaller fish and, because size is related to age, they are likely to have spawned more times than smaller fish. An expected value of SSB/R can be calculated for any level of fishing mortality (Figure 6.4). The maximum SSB/R is assumed to occur in the absence of fishing. The ratio of SSB/R for an exploited stock to the maximum potential SSB/R of that unfished stock is called the spawning potential ratio. This ratio, a measure of the impact of fishing on the potential productivity of the stock, is gaining use in U.S. marine fisheries management. Most fisheries scientists believe the spawning potential ratio should not be allowed to fall below 20%; minimum values between 20 and 30% have been recommended for conservation of U.S. groundfish stocks (10).

Historical stock–recruitment data can be used in other ways to estimate fishing mortality. First, the median ratio of SSB/R for previous years (half the ratios have been lower, half higher) is determined. Then from a graph like the

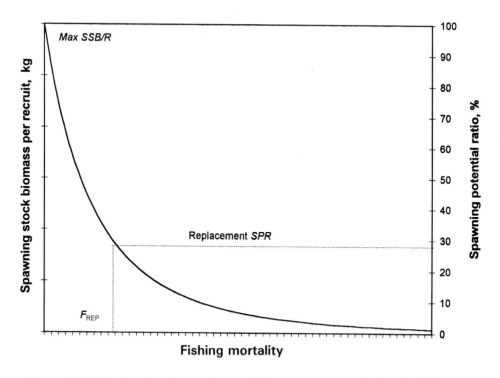

FIGURE 6.4.—Relationship of spawning stock biomass per recruit (SSB/R) to fishing mortality (modified from Mace and Sissenwine 1993). The maximum (max) of SSB/R is equivalent to 100% of the spawning potential ratio (SPR). For a representative fishing mortality rate (here, the maximum mortality that allows the stock to exactly replace itself, F_{REP}), the curve gives the corresponding values of SSB/R and SPR (follow the horizontal broken line to the vertical axes).

one in Figure 6.4, the associated fishing mortality, termed F_{MED}, is read. Theoretical studies suggest that F_{MED} is at least as reliable a target reference point as F_{MAX} and $F_{0.1}$. If the current value of SSB/R is much lower than the median SSB/R, implying that current F is dangerously greater than F_{MED}, action may be needed to sharply reduce the exploitation rate. Conversely, if this year's SSB/R ratio is well above median, restrictions on fishing might be relaxed. This use of stock–recruitment data is being explored in Europe and North America (11).

MULTISPECIES MANAGEMENT

Groundfish live in multispecies communities, and associations of particular species may be consistent over large areas (Figure 6.5). These species interact in various ways, competing for space, food, and shelter and preying on one another. Such "biological interactions" change with environmental conditions (temperature, habitat type, etc.), with the number of species present, and with the life stage of the fish. The abundance of one species affects the abundance of many other species, even in the absence of fishing.

148 Single-Species and Multispecies Management

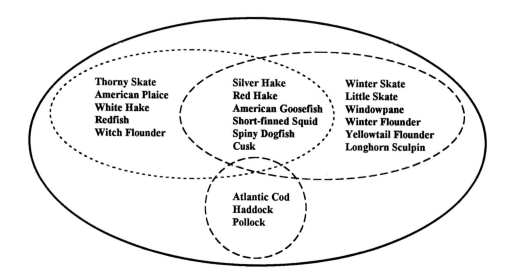

FIGURE 6.5.—Species assemblages of northwest Atlantic groundfish. Overlapping ellipses indicate how groupings shift over large spatial areas (based on Gabriel 1992).

Fishing is superimposed on this biological web, leading to "technological interactions" among species. Many species are caught with varying efficiencies by several types of gear. Conversely, the same gear may catch several species simultaneously even when only one species is targeted. For example, during an experimental fishery that targeted Atlantic cod on Georges Bank in early 1994, the proportion of cod in the catch ranged from 11 to 93% and averaged only 56%; the bycatch of nontargeted fish consisted of haddock, pollock, and nine other species (12). Different species are affected differently by fishing, and these differential effects can lead to major changes in biological interactions.

Recognition that fish are subject to myriad biological and technological interactions that affect sustainable yield has prompted scientists to consider "multispecies management" as an alternative to the traditional single-species management. The main multispecies approach tried so far consists of integrating two or more single-species models and applying optimization techniques to the new combination. The problems are formidable. Consider the simple two-species combination illustrated in Figure 6.6. The two species differ markedly in maximum yield per recruit and in the fishing mortality (F_{MAX}) that will allow the maximum yield to be achieved. Both species are caught with the trawling gear used by the fleet. If fishing mortality is at F_{MAX} (or any other reference mortality) for species 1, it is very unlikely to be at F_{MAX} for species 2; therefore, maximum yield per recruit cannot be obtained from both species at the same time. Then, management must make some compromise between the two species in order to optimize the yield per recruit of both. But how should the yield per recruit be optimized? Perhaps the two species have the same market value; in this case, managers might simply try to maximize total biological yield (weight) by setting a target F somewhere between the two values of F_{MAX} and regulating fishing effort to achieve that

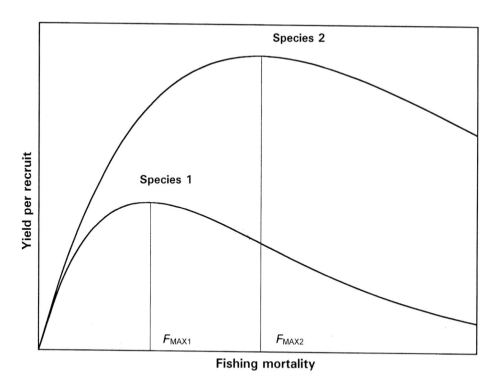

FIGURE 6.6.—Yield-per-recruit models for two hypothetical species under joint management for optimum combined yield. Maximum yield per recruit of each species is the apex of each species' curve; the corresponding fishing mortality at the maximum (F_{MAX}) is the value at the base of each vertical line from the apex.

F. Species 1 would be underfished relative to its potential and species 2 would be overfished, but the combined yield would be the greatest possible. If one species were more valuable than the other, target F could be shifted closer to the F_{MAX} of the valuable species so as to maximize gross economic yield. Further compromise might be necessary if optimizing the combined fishery would drive one species to economic (if not biological) extinction.

Multispecies management becomes rapidly more complex and expensive as additional species are brought under consideration. In practice, only a few species (or stocks) are likely to be placed under simultaneous management for fishery optimization; other species will be unmanaged, although they necessarily will be affected by controls placed on the managed group. Mathematical techniques exist for selecting the species to be managed and optimizing the overall yield. Making effective practical use of these tools requires strong management criteria for determining the species to be managed, the quantity to be optimized, and the conservation imperatives that might override normal decisions, among others.

Multispecies groundfish management has received consideration but little implementation in the northwest Atlantic (13); it is being explored most comprehensively with the North Sea fisheries of Europe, where it encompasses pelagic species as well as bottom fishes. The European approach has been to integrate single-species models into a composite multispecies model. A group of scientists under the auspices of the International Council for the Exploration of the Sea (ICES) developed a method, called multispecies virtual population analysis, that takes into account predation among exploited fish populations. The method incorporates descriptions of predator–prey relationships into traditional virtual population analysis (VPA). The latter is one of the major tools employed by fisheries biologists to estimate stock size. A VPA uses information on total catch, age structure, and natural mortality to estimate absolute abundances of fish cohorts (fish of the same age-group) and fishing mortality. In multispecies VPA, natural mortality is split into two components: mortality caused by predation and nonfishing mortality from all other sources. The total amount of a prey species (e.g., herring) eaten by a predator (e.g., cod) can be estimated each year from the abundance of the predator and the contents of predator stomachs. Losses due to predation are used to account for annual changes in prey natural mortality, and prey natural mortality then is used in virtual population analyses to estimate cohort sizes of prey species. Application of the method requires analysis of many predator stomachs, representing all predator age-groups, as often as every 3 months because predators change diets as they grow and from season to season. A strong international stomach-sampling program in North Sea fisheries began in the early 1980s. A relationship between predation mortality and catch only emerged after several years of reliable sampling, but predation now is understood to be a major source of natural mortality for several commercially important species. This insight would not have been acquired without the use of multispecies VPAs (14). Nevertheless, these models only address biological interactions; technological interactions (i.e., interactions among fisheries) still must be explored.

PROSPECTS FOR SINGLE- AND MULTISPECIES MANAGEMENT

Northwest Atlantic groundfish have always been managed as individual species. The use of mathematical models to estimate stock size, determine allowable catch, and set fishing mortality goals has occurred largely since World War II. Surplus production models first were used to estimate maximum sustainable yields of stocks and the associated F_{MSY} targets. Dynamic pool models were refined in the 1950s, and F_{MAX} supplanted F_{MSY} as a management criterion in the 1960s. Canada began implementing the more conservative $F_{0.1}$ strategy in the 1980s; the United States first adopted $F_{0.1}$ (for a few species) in 1996. Stock–recruitment relationships have been used both to estimate the recruitment expected from given stock sizes and to project the spawning stock biomass a given recruit would generate.

Single-species management is based on a comprehensive theory of fishing. For many years it seemed to be successful for northwest Atlantic groundfish in that stocks remained reasonably strong. But it was unable to prevent the collapse of many U.S. and Canadian stocks during the 1990s. Some of the fishery decline is attributable to evasion of quotas and misreporting of catches, as outlined in Chapter 3, but scientists were noting serious discrepancies between model predictions and actual events even before these collapses. Stock sizes frequently were overestimated, for example, leading to overly generous quotas and thus to fishing mortalities that reached two or three times the target levels. Reasons for these errors include incomplete data on past catches (the misreporting problem) and inaccurate estimates of fish recruitment and growth rates. Much work has been and is being done to overcome these problems, but whether it will be successful is an open question at present. If single-species fishing theory and models are ultimately discredited for groundfish fisheries, the reason probably will lie in their inability to account for biological and technological interactions among species.

Multispecies management would accommodate species interactions, and models to support this approach are becoming more realistic. Every advance in model sophistication, however, brings higher requirements for fishery and biological data. Acquiring and managing data are very expensive, and full implementation of multispecies models—as they presently are conceived—remains problematical.

Major Canadian groundfish fisheries remain closed at this writing, but the United States still allows fishing in some waters, including parts of Georges Bank. Exploitation is sharply limited, however, under terms of amendment 7 (1996) of the Northeast Multispecies Fisheries Management Plan promulgated by the New England Fisheries Management Council. The point of the limitations is to allow spawning stock biomasses of principal groundfish species to rebuild; closed areas, reductions in catch and effort, minimum fish size limits, gear modifications to reduce bycatch, and target mortalities of $F_{0.1}$ are parts of the regulatory package. Single-species management remains the norm, but it has taken a form that could lead to multispecies management. The number of managed categories has been reduced to four: Atlantic cod, haddock, yellowtail flounder, and "everything else"; the last category will be managed only to sustain current yield, low though it is. Great emphasis is being placed on controlling and monitoring bycatches and discards. Once the stocks have recovered, much of the information-gathering system needed to optimize future joint harvests may be in hand. The next step in achieving true multispecies management will be to meld knowledge of technological interactions with information about biological interactions. To reach this goal, an ecological perspective will be needed.

NOTES

1. Huxley was quoted by Smith (1994) in his history of fishery assessments during 1855–1955.
2. European work on multispecies management has focused on North Sea fisheries. As we discuss later, optimizing yields from several species simultaneously is a very demanding exercise that is viewed with skepticism as well as optimism (Daan and Sissenwine 1991).
3. A stock is a more-or-less discrete and identifiable unit of a fish species. "Stock" and "population" are interchangeable terms, but "stock" is the one usually applied to units of exploited fish species. In the strictest terms, a stock is a group of individuals that interbreed much more with each other than with individuals of other groups. In the absence of genetic tests to confirm group identity, stocks may be defined in terms of unique appearance or consistent occurrence in a particular area. In the northwest Atlan-

tic Ocean, particularly off Canada, stocks often are defined by the fishery statistical zone(s) in which they occur (e.g., the "2J+3KL stock" of Atlantic cod). "Stock" also is used in reference to part of a stock; thus "spawning stock" refers to the reproductively mature members of a stock.

4. Surplus production (and other) models are often known by their inventor's names. The Schaefer basic model was presented by Schaefer (1954). This approach was developed further by (among others) Pella and Tomlinson (1969) and Fox (1975).
5. All modern dynamic pool models are derived from the seminal work of Beverton and Holt (1957).
6. Sometimes yield per recruit does not have a dome-shaped relationship to fishing mortality. Rather, it continues to increase as fishing mortality increases until the fishery suddenly collapses. Under these conditions, F_{MAX} is a very dangerous reference point, as is F_{MSY} calculated with a surplus production model. Such a problem has not been detected for northwest Atlantic groundfish, however.
7. As shown by Deriso (1982, 1987).
8. The conservative $F_{0.1}$ strategy was suggested by Gulland and Boerema (1973). This reference point is not "optimal" for anything—$F_{0.05}$ or $F_{0.2}$ could be used instead—but it has been generally adopted as a means of controlling fishing mortality at a relatively low level.
9. Rothschild and Mullen (1985).
10. Analyses of several northwest Atlantic groundfish species have shown that recruitment declines sharply when the spawning potential ratio drops below 20% (Gabriel 1985; Gabriel et al. 1989), and this percentage is commonly adopted as a risk-aversive threshold (Goodyear 1993). Conservation goals in the range of 20–30% have been recommended for Atlantic cod, haddock, and yellowtail flounder (Anonymous 1987). Across the 91 fished stocks of 27 species examined by Mace and Sissenwine (1993), the average spawning potential ratio was 19%.
11. These ideas were introduced by Shepherd (1982) and Anonymous (1983). Jakobsen (1993) has analyzed the theoretical properties of F_{MED}.
12. Gerrior et al. (1994).
13. The inconclusive efforts to coordinate the fisheries for capelin and Atlantic cod in Newfoundland–Labrador waters (capelin are important prey of cod) and the fisheries for cod and Atlantic mackerel in the southern Gulf of St. Lawrence (young cod are prey of mackerel) were summarized by Parsons (1993). Multispecies assemblages in U.S. waters have been studied (e.g., by Overholtz and Tyler 1986 and Murawski et al. 1991), but the findings have not been applied to management.
14. The intensive efforts to develop multispecies management in the North Sea were reviewed by Pope (1991) and others in proceedings of the ICES symposium Multispecies Models Relevant to the Management of Living Resources (Daan and Sissenwine 1991).

Richard W. Langton and Richard L. Haedrich

ECOSYSTEM-BASED MANAGEMENT

The basic tools of management for North American fisheries scientists have been population dynamics models. Although ecosystem-based management cannot yet match the quantitativeness of a population dynamics approach, it does offer an alternative that addresses the interactions that operate at multiple scales throughout a fisheries ecosystem. This essay offers a definition of ecosystem-based management and suggests some necessary conditions for adopting this approach in the northwest Atlantic.

DEFINING ECOSYSTEM-BASED MANAGEMENT

Ecosystems are defined in terms of communities of plants and animals and their physical environment. They are geographically discrete units and vary in size from hundreds or thousands of square kilometers to watersheds (1), large linked systems like the Great Lakes (2), and distinctive marine areas of 200,000 km² or more (3). Ecosystems are largely self-sufficient in converting the energy they receive into the nutrients they use, although they are not biologically closed systems.

Ecosystem-based management is a new concept for resource management, and it is still being developed. The Ecological Society of America has characterized it as follows.

> Ecosystem management is management driven by explicit goals, executed by policies, protocols, and practices, and made adaptable by monitoring and research based on our best understanding of the ecological interactions and processes necessary to sustain ecosystem composition, structure, and function. (4)

In addition to that definition, the Society listed eight elements that are considered an integral part of an ecosystem-based management approach: an emphasis on sustainability over generations rather than a focus on short-term profits; well-articulated and measurable goals; use of sound ecological models; recognition of the biological diversity and complexity that allows ecosystems to adapt to long-term change; recognition that change and evolution are inherent parts of an ecosystem; realization that ecosystem processes work over a wide range of spatial and temporal scales; acknowledgment that humans are ecosystem components; and recognition that current knowledge of ecosystem functions is incomplete and subject to change.

The traditional view of ecosystems is that humans extract resources according to socioeconomic rules and that humans are not integral parts of the systems. Ecosystem-based management, with humans defined within the system, takes a perspective that the overriding rules for the governance of human behavior are ecological ones. Humans share in the exploitation of a system's resources, but the magnitude of that share must not upset the balance required to keep the ecosystem's cycles running and sustainable. This share would be based on an appreciation of multispecies relationships as opposed to the single-species focus inherent in the notion of maximum sustainable yield (5) or optimal yield (6). It is the ecological underpinnings of a fishery, not economic measures, that ultimately determine sustainability, and basic to that sustainability is a preservation of the natural capital in the ecosystem (7). Preservation of this capital will guarantee that the natural cycles in the food webs, cycles that have coevolved in the system over millennia, will continue. Ecological principles demonstrate that no system "optimum" is independent of the ecosystem itself.

RELEVANCE TO GROUNDFISH MANAGEMENT

An ecosystem approach would allow fishery management to become hypothesis driven, meaning that management strategies would be set up like scientific hypotheses and tested by research and monitoring programs. Management decisions, though still scientifically informed, would be based on predetermined goals and require consensus, or at least agreement, and commitment to an experimental protocol by both science and industry. Management "experiments" would be of a defined duration. An appropriate degree of consideration would be given to the experimental design, and methods of evaluation would be an integral part of the planning process. Implicit in this approach is the need for logical and systematic changes in management strategies as part of the development of measurable goals.

The first stage of ecosystem-based management is the coupling of ecosystem science with the formulation of management goals and the implementation of policy with well-defined objectives and strategies to achieve the goals (8). A clear understanding of human social and economic values needs to be developed concurrently. These human values must be quite explicit so that fishery managers can incorporate them into specific management strategies. A reduction of fishing effort has value as an ecological objective, for example, because the biological consequences are reasonably obvious. However, social and economic controls for overfishing have to be succinctly stated in terms of fleet composition and gear type if they are to be incorporated into a strategy for a region's management plan.

Objectives and strategies must be measurable over a specified period of time. Once a strategy is implemented as part of a plan, it should be considered an experimental protocol that will be carried out over a finite period. At the conclusion of the experiment, elements of the strategy might be changed if the desired effect had not been attained, but any changes would occur on a schedule defined as part of the management plan. For example, an experimental strategy might be an increase in the cod end mesh size of an otter trawl. An effect (if any) of the mesh size increase on a particular fish stock is expected to be measurable in 2 years. If another mesh change is made before 2 years are up, the experiment will be invalid and wasted. If the effect measured at 2 years is undesirable—or if no effect at all is detected—the strategy will be altered at that time.

ADVANTAGES

As groundfish stocks declined in the northwest Atlantic, many fishers sustained their livelihoods by turning to other resources. They have developed several new fisheries that function independently of preexisting fisheries. The potential effects of the new fisheries on established fisheries and on the recovery of groundfish stocks have not been examined, and managers presently have no effective means of doing so. For example, a recently expanded sea urchin fishery in the Gulf of Maine, being coastal, might seem independent of any groundfish fishery, but it might have an indirect effect by disrupting nursery habitats for young groundfish. An ecosystem-based management approach offers a framework for objectively evaluating potential interactions and for assessing economic gains and losses in relation to ecosystem consequences.

The single most important advantage of ecosystem-based management is its formal recognition of limits to growth. Growth in any ecosystem is constrained by a natural and finite limit, and human demands on this productivity can only be accommodated within this limit. An ecosystem-based approach to management should foster a sense of stewardship toward the marine realm such that economic expectations are incorporated into the goal of environmental sustainability.

DISADVANTAGES

Development of ecosystem-based management for the marine environment will require several levels of administration. There has been much discussion of local versus centralized control of fisheries (9), but the cumulative effects of harvesting must be monitored across local, state, provincial, and federal lines of authority for fish stocks that range widely. It is also important to understand the behavioral relationships among fishers, the fish they catch, and the prey of the harvested species. For economic reasons, fishers hunt for aggregations of fish, whether fish aggregate for feeding, breeding, or occupying limited habitat. Commercial landings data document the locations of catches, within defined statistical areas, but they do not incorporate fishers' knowledge of the fish. Such knowledge is required to manage fish stocks as they move through different jurisdictions and come under different administrative controls.

Although ecological complexity is being studied as an intellectual exercise, ecosystem-based management will require research that has direct economic consequences. Both the economic consequences of stock collapse and the complexity of ecological interactions that underlie stock sustainability will cause scientists to err on the conservative side in their harvest recommendations to managers. Such a precautionary approach will not generate scientific credibility within the fishing community. Participation by experienced fishers in the design of management experiments may ameliorate such concerns. In any case, the cost of designing, monitoring, and evaluating the experiments will not be trivial.

POTENTIAL APPLICATIONS

Examples of ecosystem-based marine management do not currently exist. We offer instead some discussion of how one might approach this management philosophy in the marine environment. We include a brief outline of a biological and sociological program of research that recognizes the multiple scales on which fish and fishers interact (10).

The first problem to be addressed is defining ecosystems in the real world. One way to do this is by applying the analytical techniques used by biogeographers, which includes the mathematical identification and mapping of recurrent groups of species or species assemblages. Groundfish survey data, which cover large areas and all species caught, could be used for this analysis. The important criterion for recognition of an assemblage is consistency in the proportional abundances of the constituent species (11). The area consistently occupied by an assemblage can be considered an ecosystem. These are regions within which ecosystem properties, such as predator–prey relationships (12), or changes resulting from general system perturbations, such as changes in water temperature (13), can be studied. Within such regions, management options such as protected areas (14), seasonal closures (15), and maximum ecosystem yields make sense.

One must go beyond recognizing patterns in fish distribution and abundance to an understanding of how those patterns were established and how they are maintained over time. An ecosystem-based research program can be daunting because the system is so complex, but there are some emergent system properties that could simplify investigations. Predator–prey relationships, for example, can be documented simply by listing the prey species eaten, or prey may be grouped by common properties such as size or life history traits (16). Such groupings reduce the analytical complexity of the system, making management options more tractable, but require an understanding of the natural history of both predators and prey, especially under changing conditions. Atlantic cod, pollock, and white hake, for example, were able to sustain and increase their populations during the 1970s despite a dramatic decline of Atlantic herring, their favorite prey, because they were able to feed on other species.

Habitat management can also be considered in functional terms. It is possible to identify habitat types that are essential for the survival of different life history stages of commercially important species (17). The gravel pavement on the northern edge of Georges Bank, for example, is essential for the survival of juvenile Atlantic cod there (18), and protection of such areas should have a positive impact on the productivity of this species. Gravel is an equally important habitat for juvenile lobsters (19), and the extent of such habitat types relative to a species' geographic range can be used to simplify ecosystem management decisions. These examples for predator–prey relationships and habitat types illustrate the shift in thinking required from the current single-species, and even multispecies, perspective of fisheries.

Implementation of ecosystem-based management is not difficult in concept but has been elusive in practice because of the sociological complexity that exists in our fisheries. Optimum yield as a strategy defined in the U.S. Fishery Conservation and Management Act of 1976 was never supposed to result in the collapse of fisheries. It was supposed to give managers a way to accommodate socioeconomic needs by temporarily allowing more aggressive harvesting, though there would be a biological price to pay. In practice, short-term economic and social "imperatives" have dominated and the biological price has been catastrophic. The same is true in Canada, where overly optimistic projections of fishery production (20), and the formal establishment of a highly technological fishing industry outstripped the ability of the stocks to reproduce themselves. Managers must realize that overly optimistic harvesting allocations have major biological effects. Short-term loans may be "ecologically acceptable," but we have currently pushed the ecosystem to an alternative state and perhaps even to a new equilibrium, particularly on Georges Bank where dogfish and skates have replaced the more marketable species (21). It is legitimate to ask if Atlantic cod, haddock, and yellowtail flounder can reestablish dominance, and if so, how fast and at what social price. If they cannot, what are the ecological factors inhibiting the fishes' recovery and what are the social consequences? Would a program of subsidized fishing to remove lower-value species hasten recovery of the ecosystem to its former equilibrium by reducing predators or competitors? Would it be biologically and economically feasible, and would it speed recovery, to enhance wild stocks with cultured fish of the appropriate size? Would a system of no-take reserves not only be socially acceptable but also enhance fish production and fishery yield by protecting specific habitats? If we assume that the ecosystem will recover (22), what level of exploitation will be allowable for sustainable fisheries in the future, and how do we allocate a limited resource in a fair and equitable manner? Answering such questions by incorporating human values into the framework of biological limits is the true challenge for ecosystem-based management of marine systems (23).

NOTES

1. Perciasepe (1994).
2. NRCUS and RSC (1985).
3. For example, large marine ecosystems as defined by Sherman (1994).
4. Christensen et al. (1996).
5. But see Larkin (1977).

6. See the U.S. Fishery Conservation and Management Act of 1976 for the formal definition of optimum yield.
7. Ludwig et al. (1993); Rosenberg et al. (1993).
8. The literature on ecosystem-based management is growing rapidly. Most papers published to date refer to terrestrial systems, but the principles they espouse are equally applicable to the marine environment. A selection of these papers include Harris et al. (1987), Slocombe (1993), Grumbine (1994), Alpert (1995), and Lackey (in press). Other references can be found in the Ecological Society of America report by Christensen et al. (1996).
9. See Wilson et al. (1994) and other chapters in this volume.
10. Langton et al. (1995).
11. Gomes et al. (1992); Haedrich and Fischer (1996). See Figure 6.5 for examples of groundfish assemblages.
12. Gomes and Haedrich (1992).
13. Gomes et al. (1995).
14. Haedrich et al. (1995).
15. Hutchings (1995).
16. Tyler (1972); Langton (1982); Langton and Watling (1990).
17. Langton et al. (1995).
18. Lough et al. (1989) and Gotceitas and Brown (1993).
19. Wahle and Steneck (1991, 1992).
20. Ommer (1995).
21. NEFSC (1995); Sherman (1992).
22. Myers et al. (1995); see also Chapter 8.
23. R. W. Langton acknowledges the financial support of the Sport Fish Restoration Act program and the state of Maine; R. L. Haedrich acknowledges the financial support of the Natural Sciences and Engineering Research Council of Canada and the Tri-Council Eco-Research Program. We also thank Johanne Fischer, Peter Sinclair, Jeff Hutchings, and an anonymous reviewer for constructive criticism of the manuscript.

Peter J. Auster and Nancy L. Shackell

8

FISHERY RESERVES

The overall purpose of designating fishery reserves is to ensure the sustainability of fish stocks and the availability of fishery resources. The additive problems of accurately assessing the status of harvested populations, making allocation decisions, and enforcing regulations regarding catch have contributed to overexploitation of many of the world's major fisheries (1). Several new ideas for fisheries management would establish risk-aversive measures to control fishing mortality. Among these, fishery reserves would allow managers to reduce mortality of juvenile and adult members of harvested populations. Fishery reserves are defined here as spatially bounded areas in which harvest of marine resources is restricted or forbidden. Such areas may be closed to all fisheries for short periods of time (weeks or months) or permanently, or they may be closed to specific types of fishing gear.

Seasonal and permanent closures have been used as conservation tools by many artisanal societies including island cultures throughout Oceania and Native Americans or First Nations peoples of North America. These societies manage relatively small areas exploited by small numbers of fishers, and fishery management is generally community based (2). Seasonal closures have also been used in large industrial fisheries, such as the groundfish fisheries of the United States and Canada, as one of several measures to reduce mortality of spawning fish as well as of juveniles (3). The value of such closures and reserves remains generally untested (4).

Reserves can be designed to protect life history stages of target species (e.g., juveniles, spawning adults), habitat features (e.g., cover from predators or substrates for spawning), or some combination of these. Harvested populations would be enhanced by emigration of fish from reserves. The issue is how to design effective reserves by considering life history characteristics of target species and identifying important habitat features (5).

POTENTIAL BENEFITS

Populations

Studies within existing coastal marine parks and reserves indicate that abundances, sizes, and reproductive outputs increased for many protected species, but not all (6). These studies have focused generally on high-topography habitats such as coral reefs, rock reefs, and kelp forests. Little information is available for low-topography habitats on temperate and boreal continental shelves. A reserve that covers a substantial portion of a species' or a stock's distribution theoretically would reduce fishing mortality of adults and the potential of recruitment overfishing. For example, a closure of the Atlantic cod fishery off southern Labrador during the peak spawning period of February to April, which was proposed in the early 1970s, would have reduced annual catch by an estimated 60% and fishing effort by 50% in the area (7). The seasonal closure was never implemented, partly because it was thought that the same total fishing effort would simply become more concentrated in other areas. In the absence of reductions in total allowable catch, such concentrated effort could have been deleterious.

Permanently closed hypothetical reserves on Georges Bank have been examined with computer models (8). Reserves could increase the spawning stock biomass of cod, but the higher the fishing mortality outside the reserves and the greater the mobility of the fish, the larger the reserves would have to be to remain effective. Reserves also allow more sedentary resources, typically invertebrates such as sea scallops, to increase in local abundance. Most such species are "broadcast spawners," casting eggs and sperm into the water for chance fertilization. When the spawners are rare, the gametes can be dispersed by currents before much fertilization occurs. A high concentration of spawners and of the gametes they release greatly increases the efficiency of reproduction. In general, reserves can be most beneficial to species whose adults aggregate for breeding but whose populations are at low levels.

Bycatches and discarding of netted fish, juveniles in particular, contribute significantly to population declines (9); 26–44% (by weight) of total groundfish catches in the U.S. Gulf of Maine groundfish fisheries were discarded in 1991 (10). Most discarded fish are injured and do not survive long. Reserves placed where juveniles congregate should reduce such losses, though the use of closures to address this problem has not been well tested. However, seasonal closure of an area in the North Sea to reduce bycatch mortalities of undersized fish resulted in an estimated 11% increase in recruitment to the sole fishery (11).

Genetic diversity may be sustained if a network of reserves can protect local stocks from extinction and preserve elements of a "natural" age structure for a species as a whole. Fisheries typically target the largest individuals available in a population. The largest fish are usually the oldest, the (then) slowest-growing, and the most difficult to replace of the population's members. As the largest individuals are fished out, attention turns to progressively smaller fish until—as has happened in many groundfish fisheries—the size of exploitable fish is at or barely above the minimum legal size (a consequence being that many sublegal fish are caught and discarded). This trend can be exemplified by the Atlantic cod fisheries. Fish become more fecund (produce more eggs or sperm) the older they become, in general. In northwest Atlantic cod populations, nearly 50% of the eggs were produced by 10–14-year-old fish in 1962. This proportion was reduced to less than 20% by 1977 and to less than 10% by 1991 (12). The genetic and other biological consequences of such profound population changes are likely to be severe, but they still are poorly understood. Although some species have shown no detectable response to fishing pressure, others produce individuals that mature at smaller sizes and younger ages, which can lead to growth overfishing. Whether or not such life history adjustments represent permanent genetic change is unknown in most cases. But substantial losses of genetic diversity were measured in orange roughy populations off New Zealand when fishing reduced the number of spawners by 70% in only 6 years, and losses in diversity are expected whenever population size is sharply lowered (13). A fishery reserve network could be designed to prevent the potential loss or change of genetic diversity by establishing protected areas over a wide geographical range. However, a reserve may not prevent a collapse of age structure without complementary changes in fishery regulations.

Habitats

A goal of marine reserves can be to protect specific habitat features. The habitat requirements of various life history stages are not currently well defined for groundfish. However, underwater observations have shown that although larval cod and haddock settle over a wide area of Georges Bank's Northeast Peak, juveniles survive best on gravel bottom (14). Laboratory studies have confirmed that gravel and cobble bottoms afford juvenile cod more protection from predators than sand bottom (15). Many other groundfish species also associate with a variety of topographic structures on otherwise nearly featureless bottom, including sand waves, burrows, and colonies of invertebrates (16).

Studies in the Gulf of Maine showed that mobile fishing gear reduced habitat complexity by destroying biological cover provided by invertebrates, reducing the numbers of burrowing species, smoothing bedforms, and dispersing aggregates of shell. Reserves permanently closed to fishing would allow habitat complexity to redevelop, enhancing survival of juvenile groundfish and potentially increasing recruitment to the fishable stock (17).

BIOLOGICAL CONSIDERATIONS

Reserve areas can be designed to protect a segment of the spawning adult population as well as to protect postlarval and juvenile life history stages (Figure 8.1). However, evaluating the effects of the closure can be difficult if fishery interactions are not taken into consideration. For example, a pot fishery for adult male red king crab occurs in the Bering Sea. A no-trawling zone was established to protect adult and juvenile crabs, but research surveys within and outside the closure zone, both before and after the closure, did not demonstrate any changes in the abundance of females or prerecruit males. The problem was that the closure zone contained no breeding and hatching areas or juvenile habitat, all of which remained vulnerable to trawling. This study demonstrated the need to consider life history characteristics of the target species when reserves are designed (18).

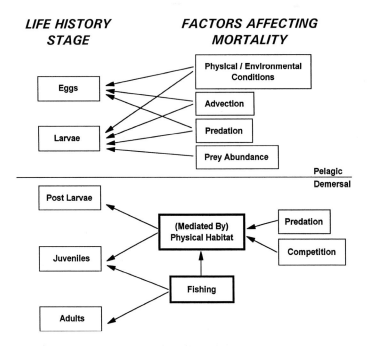

FIGURE 8.1.—Life history stages of demersal (bottom-living) fishes and the primary factors that affect their mortality rates. Advection refers to the drift of eggs and larvae with ocean currents away from settlement areas. Fishing not only contributes to direct mortality, it also reduces benthic habitat complexity. Reductions in habitat complexity may increase predation and competition, reducing recruitment to the spawning stock.

Seasonal Closures

Although closure areas can be established for assemblages of species, most seasonal closures are likely to be designed for single species or particular life history stages of a species. One species may have spatially distinct spawning, settlement, and juvenile "nursery" areas; another may occupy the same area in all life history stanzas (Figure 8.2). Designation of fishery reserves to protect some or all life history stages of a species may require that fishing be restricted in large or multiple areas. Species that form very dense aggregations at a particular life history stage would require smaller reserve areas than species with more dispersed distributions. Temporary closures can be based on annual differences in the location and success of each life history stage, though this strategy requires intense monitoring of target species and a management structure that allows rapid responses to spawning and settlement events.

Seasonal closures also can be used to protect predictable aggregations of adults and juveniles. In spring, for example, Atlantic cod move shoreward along well-defined "highways" off eastern Newfoundland and Labrador. Their routes follow relatively warm oceanic bottom water lying in trenches under cold shelf waters, and these pathways can be predicted from computer simu-

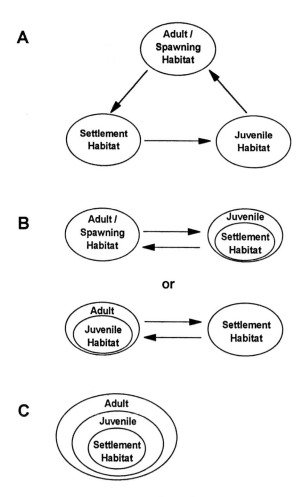

FIGURE 8.2.—Patterns of distribution in each life history stage for various demersal fish species. **(A)** Each stage occupies a distinct area. **(B)** Settled postlarvae and juvenile aggregations remain in the same habitat but adults are segregated elsewhere, or larvae settle in a specific habitat and migrate to the adult habitat as young juveniles. **(C)** All demersal life history stages occur within the same habitat. Species-specific or area-specific differences in distribution must be considered in the design of effective fishery reserves.

lations of oceanographic conditions (19). Although cod spawn over a wide area of continental shelf when they are abundant (20), dense aggregations of breeding adults occur along the spring migration corridors. Protection of such regions from fishing during migration and spawning could enhance the species' reproductive success.

Protection of groundfish nursery habitats requires knowledge of where and when larvae settle to the bottom and, for widely distributed species, where juveniles survive best (21). When fish abundance and survival vary widely from place to place and year to year, seasonal closures are more likely to achieve management goals than permanent closures.

Permanent Closures

Year-round nonextractive reserves can be designated for single species but they would be most useful for multiple species. Reserves that transcend or are within defined species assemblages (Figure 8.3) would maintain species interactions. Groundfish distributions are correlated with depth and temperature, and assemblages of species are quite stable in the absence of heavy fishing or major oceanographic change (22).

Within the general ranges of depth and temperature preferred by groundfish species, local abundances of fish are strongly influenced by the distribution of suitable bottom habitat. The importance of habitat type is known in general, but the features of habitat that best support particular species at particular times re-

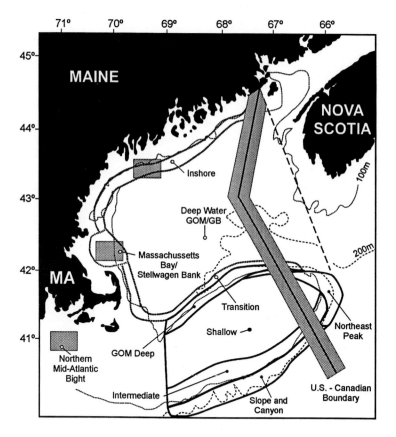

FIGURE 8.3.—Groundfish assemblages identified off the northeastern United States, named for the depth or geography of their occurrences. Solid black lines enclose assemblages associated with Georges Bank; heavily dotted lines delimit other continental shelf assemblages (data from Overholtz and Tyler 1985 and Gabriel 1992). The shaded area around the U.S.–Canadian boundary was proposed by McGarvey and Willison (1995) as a nonextractive fishery reserve. Additional shaded areas represent possible reserves to include nearshore assemblage types. The heavy broken line is a former international boundary; lighter broken lines show the 100-m and 200-m depth contours. The abbreviations GOM and GB refer to the Gulf of Maine and Georges Bank.

main poorly understood for groundfish. As a starting point for permanent reserves, published scientific information and the practical knowledge of fishers could be used to define assemblage areas containing habitat types ranging from mud through boulders. Reserves then could be refined in size as knowledge about habitat requirements increases (23). The technology of modern geographic information systems will facilitate such refinements.

MANAGEMENT CONSIDERATIONS

Fisheries management is inexact because complete information about complex systems is unaffordable and partial information obtained from sampling and research can be inaccurate or misleading. Fishery reserves offer a buffer against uncertainty. A system of no-, limited-, and full-fishing zones can become a cornerstone of fisheries management (24).

The justification for fishery closures lies in long-term sustainability of harvested populations. In the short term, however, closures have economic and social costs. Because the utility of reserves has yet to be demonstrated in practice, anticipated costs may inspire considerable opposition to a reserve system.

Enforcement may be one of the greatest impediments to successfully implementing a system of reserves. Restrictions on fishing gear can be verified by routine vessel inspection, but respect for reserve boundaries can only be verified with constant monitoring of fishing vessels. Current levels of patrolling with ships and aircraft allow a very small percentage of vessels to be monitored on any given day, and full compliance probably can be assured only when an electronic vessel tracking system is established (25).

Both acceptance and enforcement of fishery reserves will be enhanced if fisher communities are persuaded that reserves are beneficial management tools and if the communities have a strong role in the design and siting of reserves. Managers, scientists, and fishers could work towards socially acceptable reserve areas based on the principles of conservation.

CONSEQUENCES OF FISHERY RESERVES

The use of fishery reserves as a management tool should be approached as an experiment (26). The program should include research to determine the effectiveness of reserves in enhancing populations and protecting habitat—research that can account for the effects of external management measures. As much as possible, reserves should be designed around the life history characteristics of key taxa. A poorly designed reserve system can provide a false sense of security while accruing no net benefit for conservation of an exploited population (27).

NOTES

1. FAO (1994).
2. The use of reserves by indigenous cultures, in North America and beyond, has been addressed by Johannes (1978), Berkes (1987), and Wilson et al. (1994).

3. For example, NEFMC (1977, 1993) for the northwest Atlantic groundfish complex; IPHC (1987) for Pacific halibut; and ICNAF (1974), Fanning et al. (1987), and Halliday (1987) for haddock.
4. ICES (1994) and Man et al. (1995), but see Campbell and Pezzack (1986) for American lobster, Klima et al. (1986) for shrimp, Rice et al. (1989) for bivalves, Bennett and Atwood (1991) for surf zone fish, and Armstrong et al. (1993) for king crab.
5. Langton et al. (1996).
6. Roberts and Polunin (1991); Bohnsack (1992); Carr and Reed (1993); Dugan and Davis (1993).
7. Wells and Pinhorn (1974).
8. Polachek (1990); Quinn et al. (1993); McGarvey and Willison (1995).
9. Alverson et al. (1994).
10. Murawski (1993b).
11. Rijnsdorp and van Beek (1991).
12. Hutchings (1995).
13. Orange roughy genetics were presented by Smith et al. (1991). The effects of fishing (or their absence) on the life history and genetics of other species have been discussed by (among others) Ricker (1981), Allendorf et al. (1987), Ryman and Utter (1987), Plan Development Team (1990), Rijnsdorp et al. (1991), Hanson and Chouinard (1992), Hilborn and Walters (1992), Policansky (1993), and Rijnsdorp (1993).
14. Lough et al. (1989).
15. Gotceitas et al. (1993); Gotceitas and Brown (1994).
16. Auster et al. (1991, 1995, 1997); Malatesta et al. (1994).
17. Auster and Malatesta (1995); Auster et al. (1996).
18. Armstrong et al. (1993).
19. Rose (1993).
20. Hutchings et al. (1993).
21. See Fanning et al. (1987), Lough et al. (1987), Walsh (1992), and Walsh et al. (1995), among others.
22. Overholtz and Tyler (1985); Mahon and Smith (1989); Gabriel (1992); Gomes et al. (1992); Perry and Smith (1994); Haedrich et al. (1995).
23. Langton et al. (1995).
24. Hilborn and Walters (1992); Agardy (1995).
25. NEFMC (1993).
26. CDFG (1993); SAFMC (1993); ICES (1994); Auster and Malatesta (1995).
27. P. J. Auster was supported by the National Undersea Research Center for the North Atlantic and Great Lakes. N. L. Shackell was supported by Dalhousie University. Our views do not necessarily reflect those of any government agency. Jim Rollins and Carol Lee Donaldson expertly manipulated electrons to produce Figure 8.3.

Joseph T. DeAlteris and Dana L. Morse

FISHING GEAR MANAGEMENT

Technological developments of the last half century have largely tamed the ocean wilderness. Many fishery resources that once seemed inexhaustible have been depleted by an unprecedented increase in fishing efficiency.

At the start of the Twentieth Century, when offshore fishing was primarily done from sailing vessels, groundfish in the northwest Atlantic Ocean were captured with small trawls or with hooks and lines. Hooks either were fished singly or fastened in large numbers along a single "ground" line, an arrangement called a longline (Figure 9.1). Atlantic cod, haddock, and flounders were the principal target species then, as now. Fleets of dories with hook-and-line gear set out from a single sailing mother ship and returned to the ship at the end of each day. The harvest efficiency of individual fishers was relatively low, but groundfish stocks were abundant, thus overall landings were high.

Towed fishing gear was developed during the years of sail. The first was the beam trawl, a large funnel-shaped net whose mouth was held open by a steel frame, or beam (Figure 9.2). By towing a net behind a moving vessel, fishers could cover more ground than static hook-and-line gear allowed and therefore could catch more fish. Nets also retained more of the fish they encountered than hooks did. Thus, nets were a more effective harvesting device than hooks and lines. But fishers were limited by the size of the beam trawl they could pull over the bottom under sail, in the area they could fish, by the properties of the netting that was available, and by the difficulty of retrieving bulky gear by hand.

Incorporation of steam engines (and later internal combustion engines) in fishing vessels reduced the limitations related to power, and this stimulated major developments in gear technology. Powered vessels could tow larger beam trawls, and powered winches could retrieve the gear more easily. Large beam trawls, however, are very unwieldy to handle and limited in bottom coverage. Fishing gear engineers discovered that small flat boards (otter doors) attached to

168   Fishing Gear Management

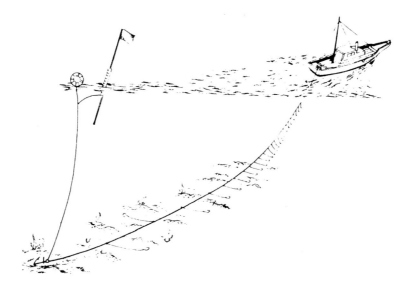

FIGURE 9.1.—Longlining configuration typical of groundfish operations in the northwest Atlantic Ocean.

tow ropes near the sides of the net could open the net mouth more effectively than the bulky frame, which led to development of the modern otter trawl (Figure 9.3). Being lighter, otter trawls could be "flown" above the bottom (unless they were weighted to stay down), which allowed trawling over areas of rough bottom for the first time. Indeed, depending on the power and thus speed of the towing vessel in combination with the size of the net, otter trawls could be pulled at any height off the bottom, giving rise to "midwater" trawls. Engineers and fishers soon also discovered that the otter doors could be separated from the wing ends

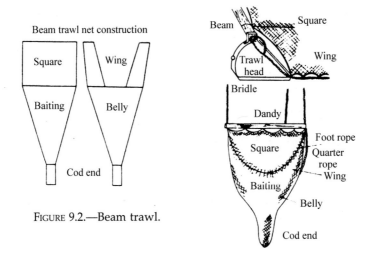

FIGURE 9.2.—Beam trawl.

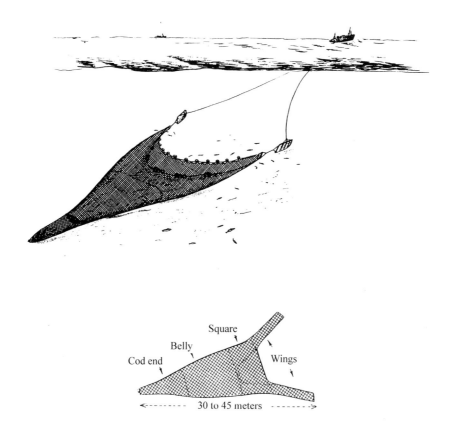

FIGURE 9.3.—Typical setup for an otter trawl.

of the trawl, increasing the effective fishing area from which fish were herded into the net mouth. The combination of increased vessel power and improvements in the design of the otter trawl led to the use of ever-larger nets, capable of retaining tons of fish.

The late 1950s ushered in an era of further technological improvements. Chief among these were advances in fish detection devices and electronic navigation, both of which benefited strongly from prior military research, and in the materials and hardware used in fishing gear.

Fishers once had to rely solely on their own knowledge of the habits of fish to capture them effectively. Test tows were necessarily common, though time-consuming. Now echo sounders (fish finders) helped to remove the risk of making an empty tow or of losing the gear entirely. Modern fish finders include false-color monitors and computer displays of precisely where the fish are, what species they are likely to be, and exactly what the contours of the seabed are below and ahead of the vessel (1).

Navigation greatly improved after World War II with the introduction of loran (LOng RAnge Navigation), which allowed fishers to relocate areas of previously abundant fish (2). Today, the global positioning system (GPS) can continuously locate a vessel anywhere to within 10 m. Fishers now can retrace any course precisely with the help of computer-driven systems that display bottom contour, water temperature, course and speed, fish in the water column, and a host of other details that make them highly efficient hunters. Good radio communications among vessels further help focus fishing effort on the greatest concentrations of fish.

New materials have played important roles in the successful harvest of groundfish, particularly in gill-net (Figure 9.4) and longline fisheries. Gill nets became more effective because monofilament netting (introduced in the 1960s) is virtually invisible in the water. Fish are not easily able to detect monofilament nets, so they are likely to become gilled or entangled in the twine. Fish had less trouble avoiding the more visible cotton, manila, and sisal nets that had been used previously. Longlines also became more effective in catching groundfish when they exploited the strength and invisibility of monofilament for leaders and more modern hook designs.

Other major areas of advancement are in evaluations of gear performance and fish behavior. Engineers and behaviorists use underwater cameras, flume tanks, and scale models to better understand the capture process. Scale models tested in the laboratory give cost-effective insights into the responses of gear to various hydrodynamic conditions. In the field, cameras and other types of acoustic net-monitoring systems provide data on the spread of trawl doors and wings, the height of the headrope, the amount of catch and the species caught, and the ways fish react to the gear (3). By better understanding the behavior of fish in general, engineers and fishers can develop more effective fishing gears and harvesting strategies.

The overall effect of the advances in fishing gear and related technologies has been to make the fishers more efficient, allowing them to catch, store, and transport more fish over longer distances in a shorter time. Fishers now can venture into areas once viewed as unfishable, can explore deeper water and rougher seabeds, and can withstand stronger currents and larger waves. For fishers, and thus for fish, the ocean has truly become a smaller place.

ECOSYSTEM EFFECTS OF HARVESTING

Advances in fishing technology and efficiency during the past 40 years have had profound consequences for marine ecosystems (4). They have contributed to fish stock depletions, shifts in species abundances, and habitat degradation. In future efforts to rebuild and then sustain groundfish fisheries in the northwest Atlantic, gear management will be of major importance.

Some ecosystem consequences have arisen because fishers are selective in their harvests. They are selective both because they are in business and have to obey market forces to make a living and because they are constrained by fishery regulations. In general, fishers attempt to harvest one or perhaps two "target" species. In the process, they inevitably catch other "nontarget" species, often referred to as "bycatch." The bycatch may be very important to fishers; sometimes it is worth more than the target species. Some of the catch—of both targeted and

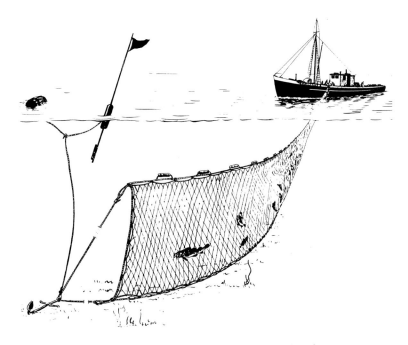

FIGURE 9.4.—Gill net, as used in groundfish fisheries.

nontargeted species—may end up as "discards," individuals thrown back to the sea because they are below legal size or age limits, because they exceed catch limits, or because they have no market value. Most fish are injured to some degree by the capture process, so discards typically have extremely high mortality rates from predation and stress-induced disease, if not from the injury itself (5). Thus, discarded fish are wasted because they will not be available for future harvests, and fisher time and energy have been expended to bring them on board and then abandon them. In addition, nonreporting of discards causes managers to underestimate fishing mortality, especially for juveniles, and thus to allow higher catches than a stock can sustain. For reasons both obvious and hidden, it benefits fish and fishers to bring up only what will be used.

Fish are removed somewhat selectively from wild populations. One effect is the removal of the largest individuals, which typically exceed any gear (e.g., mesh size) regulations that may be in effect; large fish usually bring the highest market price as well. But these individuals contribute greatly to the reproductive potential of the species, and their loss often affects future stock size and recruitment. Another effect of selective harvesting is a shift in biomass from one species or group of species to another. For example, scientific monitoring has shown a marked change in the biomass composition on Georges Bank over the last 30 years from cod, haddock, and yellowtail flounder to sharks and skates (6). The Georges Bank system now produces species that are less profitable in the seafood market than the original species, but fishing still is being driven by the consumer market, a factor that has been only recently appreciated.

Fishing activities also disturb the seabed. Trawls, scallop drags, and hydraulic dredges for shellfish disrupt or destroy animal or plant life occupying the upper layer of the sea floor, down to several centimeters (7). Longlines, traps, pots, and gill nets also affect the physical environment, though to a smaller degree. Habitat changes affect not only the survival of adult groundfish but also (and usually more importantly) the survival of larval and juvenile stages of these species and of the invertebrates on which they feed. Gear-induced changes in bottom characteristics appear to be important and well worth more thorough investigation.

Thus, fishing activity has (and has had) major influences on wild fish stocks. To a great extent, these effects are caused by the characteristics of the gears used to harvest fish. A major task facing today's fisheries managers, biologists, conservation engineers, and fishers is to develop gears that allow fisheries to be economically viable but that are selective in appropriate ways and that impose minimum impacts on physical and biological environments.

SELECTIVITY

Options for improving the selectivity of harvesting gear to maximize the catch of target species and reduce inappropriate bycatch fall into three broad categories, representing selection by (i) time and place, (ii) fish size, and (iii) species. One degree of selectivity is achieved by simple decisions about where and when to set gear. If an area is deemed too rocky to fish, has an excessive number of spawning or juvenile individuals of a species, or is dominated by another gear type, a fisher may avoid it. Managers can also impose this kind of selectivity by specifying time and area closures.

Selection by size or species is more problematic. Behavioral differences within species (e.g., between juveniles and adults) and between species (e.g., between flatfishes and roundfishes) must be considered. So must the co-occurrence of important groundfish species in time and space. Gear set by a fisher is likely to interact with many species. This reality, combined with selective properties inherent in the gear (or the lack thereof), creates a difficult situation for managers, fishers, and scientists trying to walk a fine line between minimizing bycatch and discards and securing a measure of fishing profitability (8).

The size selectivities of gears most commonly used in the northwest Atlantic groundfish fishery are broadly understood. For trawls, the cod-end mesh size is most important, because most fish escape the net there (9). Diamond-shaped meshes (in nets "hung" with a mesh corner upwards) favor escapement of flatfishes, but escapement is limited to meshes immediately forward of the fish mass, because more forward meshes are pulled closed. Cod-end meshes hung square (the top of a mesh is parallel with the towed net) favor escapement of roundfishes. With this knowledge, gear designers can maximize selectivity with a minimum reduction in the value of a fisher's catch.

Gill nets are more size-selective than trawls, being most efficient within a narrow band of fish sizes (10). Below the optimal size range, most fish pass through the meshes of the net; above this range, most individuals "bounce" off the twine without becoming entangled.

Selection by species requires an understanding of fish behavior, a complex subject. Species may react similarly to a gear at some point in the capture process but differently at another. When otter trawls are dragged on the bottom, for example, the doors raise sand clouds that herd both Atlantic cod and haddock to a station slightly off the seabed and between the wing ends of a trawl. As they tire, cod turn back into the trawl and dive, whereas haddock try to rise over the headrope (11). So, a fisher aiming to reduce the catch of haddock in cod tows will modify the net, reducing headrope height, to allow haddock to escape more easily.

Some practical examples illustrate how gear research and modification can reduce bycatch. The trawl fishery for silver hake in New England is conducted with small-mesh (7.5-cm) nets that often make high incidental catches of flatfishes, most species of which are highly regulated. Observations and initial trials confirmed that flatfishes and hake enter the trawl mouth at different heights above the seabed, the flatfishes within 0.6 m of the bottom and the slender-bodied hake and other roundfish some distance above that. A series of cooperative studies involving commercial fishers, state managers, and scientists led to the addition of dropper chains that kept the lower mouth of the trawl off the bottom, so fewer flatfish would enter the net, and incorporation of a large-mesh panel in the lower trawl belly facilitated escape of the flatfish that did enter. These changes reduced the bycatch of flatfish from more than 10% to less than 5% of the overall catch by weight (12).

The New England trawl fishery for northern shrimp also deploys small-mesh nets, and it has a bycatch problem with juvenile groundfish. One solution exploits differences in swimming ability and body shape between shrimp and fish. A rigid Nordmore grate with a 2.5-cm bar spacing is installed into the extension section of the trawl (Figure 9.5). Shrimp, which are small and poor swimmers, are swept through the openings of the grate. Finfish stopped by the grate are able to swim against the current for a while and reach an escape vent above the grate (13). Their escape is eased if the fishing vessel is slowed periodically to relieve fish pinned against the grate, one of many cases in which the skill and motivation of the vessel captain can make gear modifications more effective at reducing bycatch.

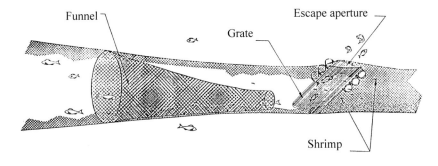

FIGURE 9.5.—The Nordmore grate system as used in a shrimp trawl.

A final example of enhanced species selectivity involves the use of sound. In the northwest Atlantic, seals, porpoises, and whales sometimes are entangled and killed in gill nets that are set on the bottom for cod and flatfish; considerable damage to the gear can result as well. These marine mammals respond to and learn from sound. When acoustic "pingers"—devices that produce sound at a particular frequency—are attached to gill nets, marine mammals have shown a strong tendency to avoid the gear. (14).

FUTURE OF GEAR MANAGEMENT

Advances in fishing gear design during the past 100 years have brought so much capture efficiency that most commercial fisheries now are overexploited. Further improvements in fishing vessels, gears, and techniques are possible and probably will be made to gain economic efficiency. However, increases in bulk fishing power will have to be offset by reductions in the amount of fishing effort if commercial fish stocks are to be sustained at viable levels.

As implied earlier in this presentation and elsewhere in this book, fishing gear management will be increasingly influenced by the biology of the species hunted. Gear engineering will aim primarily at reducing bycatch (improving species selectivity) and minimizing habitat damage. Time and area closures for some or all gear will be used to protect particular life stages of fish (such as juveniles and migrants) and the key habitats on which they depend. The success of such management will depend on detailed knowledge of the life history, behavior, and ecology of exploited species.

The future of fishing gear management in northwest Atlantic groundfish fisheries is being driven by the priorities of rebuilding stocks and of habitat protection. The present harvesting technology, including bottom trawls, gill nets, and longlines, all have bycatch problems, habitat impact problems, or both. Fish traps, presently used in the northwest Atlantic mainly to harvest cod off Newfoundland and Labrador but used extensively in bottom fisheries elsewhere in the world, offer many potential benefits in terms of selectivity, discard survival, and minimal habitat impact. With cooperation among fishers, scientists, and resource managers, continued progress in the development of optimal harvesting technologies can be achieved.

NOTES

1. MacLennan and Simmonds (1994) provided a comprehensive overview of recent advances in fish detection.
2. Loran-A, which came into use during the 1950s, determined position to within 1,000 m. In the 1970s, loran-C gave position to within 100 m. The operation of these and more modern navigation systems has been described by Bowditch (1995).
3. Net-monitoring systems based on acoustic transducers are available to both fishers and fishing gear technologists.
4. Dayton et al. (1995) described these effects in more detail. See also Chapter 8 in this book.
5. High mortality of discarded northwest Atlantic groundfish has been documented by Howell and Langan (1987) and DeAlteris and Reifsteck (1993).

6. This trend is documented in Anonymous (1993).
7. See Jones (1992) for an overview of trawling damage to seabeds.
8. Multispecies catches also complicate uses of fishery models (Chapter 6).
9. See Robertson (1988), Robertson and Ferro (1988), and DeAlteris et al. (1990).
10. Lazar (1993).
11. Wardle (1986) and Nunallee (1991) have documented this behavior.
12. Early work on flatfish bycatch in the silver hake fishery was done by Carr and Caruso (1992). DeAlteris et al. (in press) designed the current trawl modifications.
13. Kenney et al. (1992) described the success of Nordmore grates in New England shrimp trawls.
14. The application of pingers to gill nets was modeled on successful efforts to keep whales away from cod traps (Lien et al. 1990).

RALPH E. TOWNSEND AND ANTHONY T. CHARLES

USER RIGHTS IN FISHING

Who is entitled to take part in a fishery? This question focuses on the issue of "user rights," a theme with a lengthy history in fisheries. Indigenous peoples have long determined the use of local resources through "customary marine tenure" (CMT) and "territorial use rights in fishing" (TURFs). In modern state management of fisheries, user rights were initially created by giving a limited number of individuals the right to fish, which was called "limited entry." More recently, user rights have defined either a right to fish within specified limits on fishing effort (such as a maximum vessel size, some limited amount of gear, or a limited number of fishing days) or a right to catch a specified amount of one or more species of fish.

User rights must be distinguished from owner rights. The state (or the community in the case of CMT and TURFs) retains the rights and responsibilities of an owner. For example, the government decides the best conservation strategies, rather than giving that responsibility to individual fishers. User rights generate some, but not all, of the incentives for efficient resource use that would occur under strictly private ownership. How well these incentives function depends in large part upon how well the user rights are defined.

CUSTOMARY MARINE TENURE AND TERRITORIAL USE RIGHTS

User rights have their longest history in indigenous fisheries. For example, in the Maritimes region of Atlantic Canada, the Mi'kmaq have developed a social process for determining control over fishing territory.

> In the centuries before the arrival of the first Europeans, the Mi'kmaq...governed themselves through councils based on consensus in accordance with the laws of nature. District Chiefs were responsible...for confirming and reassigning hunting/harvesting territories. (1)

This represents a form of territorial use rights in fishing, or more generally, customary marine tenure. Such systems assign to individuals and to groups the rights to fish in certain locations, generally, although not necessarily, based on long-standing tradition ("customary usage"). These systems, which can provide relatively stable, socially supported tools of fishery management, are more prevalent globally than is reflected in the fishery literature (2).

Although TURFs and CMTs are best documented in the context of developing countries, notably in the South Pacific, they are found throughout the world. Lobster fishers in some areas of Maine have been able to maintain informal, nonlegislated control on entry, for example (3). Along northeastern Cape Breton Island, Nova Scotia, a lobster poaching problem in the late 1800s was resolved by the local Minister, who decreed marine use rights based on an extension of property lines out to sea. The resulting management system has been maintained by the community to this day (4). In another example of informal use rights in Nova Scotia, applying in this case to fisheries for groundfish and herring as well as for lobster, "Claims of ownership and control of property are centered in the community, and individual use-rights are derived from membership in the community" (5).

Existing CMT and TURF systems may provide efficient means of fishery management, but they tend to be poorly understood and neglected by governments. There may be value in nurturing such approaches, where they exist, and in examining the potential for their development in other fisheries (6).

LIMITED ENTRY

Under limited entry, the government issues a limited number of licenses to fish. This creates a user right—the right to participate in the fishery. Limited entry prevents the entry of new fishing boats and hence tends to limit the overall amount of fishing effort. If limited entry is successful, this reduction in fishing effort helps to conserve the resource and also generates higher incomes for the permit holders. Although limited entry has rarely been used in the U.S. waters of the northwest Atlantic, essentially all significant fisheries in Atlantic Canada are subject to limited entry.

Limited entry has produced some economic benefits. In several fisheries, limited entry slowed the rate of expansion of fishing capacity during the 1960s, 1970s, and 1980s, when large increases in fish prices would otherwise have attracted many new entrants. License holders often earned substantial profits, which were reflected in the high market value of transferable limited entry permits.

Some of the most successful limited entry programs have been in Alaska, where the state government actually reduced active permits. Permits have retained very high values in many Alaskan fisheries. The salmon fishery of British Columbia is often cited as a limited entry program that initially created economic benefits, but whose success was eroded by the gradual increase in fishing power.

There have been two key problems with limited entry. First, most limited entry programs began as moratoria on entry after the number of participants in the fishery had already become large. Only in rare cases has government reduced the number of licenses either by revoking or buying back licenses. Second, each vessel has an incentive to expand its fishing capacity, a problem known as "capital stuffing." As a result, limited entry alone has almost never been successful in the long term at preventing economic overfishing (7). Because of these limitations, there has been an evolution in fisheries management away from simple limited entry towards allocations of individual rights to use gear or of individual rights to harvest a share of quotas.

GEAR USE RIGHTS

Typically under limited entry, too many permits were allocated and each of the overabundant fishers expanded his or her own use of fishing gear. An obvious solution to this overcapitalization would be to limit the use of some components of fishing effort, such as vessel size and fishing days. For example, limits on the number of traps individual fishers can use are long established in some lobster fisheries. Such restrictions would create a value not for the permit to fish but rather for the right to use each unit of gear.

The difficult problem for a gear use rights program is to define an adequate index of fishing activity, which must be based upon some key components of fishing effort. No matter how the index is defined, each fisher will try to manipulate both the regulated and unregulated components of fishing effort to achieve greater harvest from the allowed fishing gear. Thus, if boat tonnage is used as the index, the vessel owner will use larger engines, more electronics, and a larger crew and may redesign the vessel to increase hold capacity. Incentives are also created to fish the vessel more days per year by reducing time at the dock, by using rotating crews, and by fishing in poorer weather. Technological improvements also gradually increase the fishing effectiveness of any set of fishing gear.

However, recognizing the incentives for manipulating the rules does not necessarily invalidate the approach. Adequate indices of fishing activity (i.e., gear use) may be available. For example, in lobster fisheries in Atlantic Canada, the number of lobster pots seems to have been a relatively good index of fishing pressure. This index might be further refined by using the number of pot hauls or trap-days as an index. In trawling fisheries, an index that is based upon hold tonnage, total vessel horsepower, and days fished may be adequate.

An individual gear use program can also adjust to improvements in fishing efficiency by periodically "taxing" the inputs. For example, if fishing power per unit of fishing gear increases at 2% per year because of capital stuffing or technological improvements, this can be offset by reducing gear allocations by 2% each year. Another option would be to reduce gear allocations when a vessel is replaced. Because this reduction would occur relatively infrequently, a relatively large reduction (such as 10 or 15%) would be required. Consider, for example, a 10% replacement tax in a system of gear use rights with fishing activity indexed to vessel tonnage. To replace a 100-ton boat, the builder would be required to have 110 tons of gear-use rights; 10 tons would be permanently retired when the vessel entered the fishery.

The lobster fishery of Atlantic Canada has reduced the number of traps in various regulatory areas by lowering the maximum number of traps per boat. The economic evidence suggests that this policy has improved the welfare of fishers, although perhaps slowly (8). In the United States, the lobster industry is now discussing management plans that include trap limits that are reduced over time.

The allocation of gear use rights can be an effective form of rights-based management if care is taken in the definition of the index and if a plan to deal with improvements in fishing efficiency is incorporated. The major advantage of systems of gear use rights over individual harvest quotas (discussed next) is that enforcement costs may be significantly lower for gear use allocations. Systems of gear use rights may also be politically more acceptable to some groups. Better fishers, who can earn more income per set of their gear, often feel that their skill is better rewarded under gear use allocations than under harvest allocations.

HARVEST RIGHTS

Individual harvest rights allocate segments of a total allowable catch (TAC) to individual fishers. By allocating harvest rights, incentives are created for each fisher to catch his or her allotted fish at the lowest cost. The incentives for capital stuffing that have undermined limited entry and gear use allocation programs are largely eliminated. Empirical evidence confirms this; fishing efficiency has risen dramatically when individual quota systems have been implemented. This may be most pronounced when the harvest rights are transferable among fishers, as with "individual transferable quotas" (ITQs). The latter have become the management approach of choice for many fishery economists (9).

Atlantic Canada has considerable experience with various forms of individual quotas. In an early program, the two largest operators of trawler fleets in Atlantic Canada were given "enterprise allocations" that allowed each firm to decide how to catch its own allocation. Although these were not transferable, the program did lead to rationalization of effort within these industrial fleets (10). More recently, the individual quota program for the inshore southwest Nova Scotia groundfish fleet was implemented in response to substantial reductions in total allowable catches and correspondingly dramatic declines in the aggregate level of fishing activity (11). The consolidation of effort that occurred under this individual quota program (which allows limited transferability) is credited by those remaining in the fishery with enabling them to survive the stock and quota declines—a point supported by examination of the prices paid to obtain quota.

Individual quota programs are also found in several smaller fisheries in Atlantic Canada and elsewhere in Canada. Some are based on ITQ systems (e.g., the Atlantic Canadian herring fishery), some on nontransferable systems (e.g., the inshore groundfish fishery in the northern Gulf of St. Lawrence), and others on community-based quotas (e.g., in the Arctic) (12).

Although individual quota programs have produced some clear economic benefits, the impact on sustainability of the resource is less clear, due largely to the major difficulties in enforcement of such programs. One such difficulty arises because the annual lease value of a unit of quota often exceeds 50% of the landed value. That is, fishers often pay $0.50 for the right to land $1.00 of fish. When landings are not reported, this $0.50 lease value is appropriated by the violator. This creates a very large financial incentive to underreport landings, and an enforcement system must keep such cheating to a very low level.

Credible enforcement requires third-party counting of all landings. It is not sufficient to monitor dealer reports; collusion by dealers and fishers has undermined many quota systems. Enforcement is easiest if a few boats land at a few harbors and sell to a few sellers. Enforcement becomes more complicated if there are many boats, many harbors, and many buyers. In multispecies fleets, enforcement can also be complicated by misreporting of species. Enforcement is particularly a problem if fishers can easily sell fish directly to final consumers or to restaurants. For example, monitoring the sale of lobsters in Maine in the summer would be essentially impossible because many fishers sell "over-the-side" in small lots to tourists and small restaurants.

A second threat to conservation, and an even more challenging enforcement problem, exists at sea. The potential for increases in discards due to high-grading and dumping under individual quota systems has captured considerable attention. When the price differential between large fish and small fish is large, a fisher with an individual quota may have an incentive to discard small, though marketable, fish in order to use the quota for larger, more valuable fish. This incentive depends on the costs incurred in discarding and subsequent harvesting. If a small fish is worth $1.00 per kilogram and a large fish is worth $1.50 per kilogram, discarding the small fish is profitable for the fisher if the additional cost of catching the large fish is less than $0.50 per kilogram. In multispecies fisheries, ITQ fishers may also dump incidental catches of fish for which they cannot obtain quota.

These are major concerns in the multispecies groundfish fisheries of the northwest Atlantic, where dumping and high-grading have contributed to the recent stock collapse (13). Short of around-the-clock observer coverage on all fishing vessels, no solution is evident to deal with this conservation problem.

THE ALLOCATION PROBLEM

In a rights-based program, the rights must be allocated. There may be good reasons to auction the rights, but political realities have been that rights are allocated to fishers on the basis of historical participation. Quite simply, this allocation problem is messy; there is no "right" way to allocate rights. Predictably, those who feel short-changed by the allocation process will oppose the entire program.

A seemingly endless set of arguments can be made about how best to define "historical participation." If only the recent history is used in the allocation, fishers who exited the fishery because stock conditions were poor are penalized. If recent history is not used, recent entrants (who are often numerically large and politically important) are penalized. When there are differentiated subfleets, such as gillnetters and trawlers, the allocation problem is complicated by the divergent interests of each fleet. Elaborate weighting schemes often are developed, which may have the effect of giving everyone about the same allocation.

Allocation problems are posed by fishers who had boats under construction, whose boat sank, whose records burned, who had a serious illness, who were called to active military duty, or who died during the qualifying period. Some kind of quasijudicial appeals process for managing all these special cases is inevitable. It is important to design an appeals process that creates incentives for other fishers to challenge false information. If the total number of rights is determined in advance, fishers with legitimate claims will have an incentive to challenge false claims of participation so their share will not be diluted. If the rights will expand to accommodate new claimants, fishers may have no incentives to challenge false claims (14).

TRANSFERABILITY OF RIGHTS

When the initial rights are allocated under rights-based management, a key policy issue is often whether or not to make the rights transferable. Two approaches to transferability are possible: nondivisible transfer of fishing permits with any attached individual rights for gear use or individual harvest quotas; and divisible transfer of gear use or harvest rights. Under the first approach, the license to fish, along with all attached gear use or harvest rights, could be transferred. Under the second approach, the gear use or harvest rights could be divisible and freely transferable from one vessel to another. Subsidiary issues are whether or not to permit mortgages against the right and to limit the total acquisitions by any one firm.

Economists tend to support complete divisibility and unrestricted transferability of gear use rights or harvest rights. Transferability allows gear use rights or harvest rights to be bought and sold as a commodity, thereby maximizing efficiency in production. Fishers that sell their rights are compensated for leaving the fishery.

All fisheries management regimes must eventually deal with transferability of permits for the simple reason that fishers eventually die. If a management plan initially opts for nontransferability, hoping to reduce fishing power through attrition, fishers generally try to retain their permits as long as possible in case the decision is later made to allow transferability. Attrition may be slow and often inequitable, achieved largely by penalizing the survivors of fishers who have the misfortune of dying too soon. If the nontransferable permit applies to the vessel, incentives are created to fish the boat beyond its technological life, thus creating safety problems. And adding a "use-it-or-lose-it" provision just forces continued effort by fishers who would otherwise have left the fishery temporarily, further increasing fishing pressure.

However, unrestricted transferability generally leads to concentration of fishing rights. There are legitimate social and economic reasons to be concerned about the concentration of rights ownership. For small-scale fisheries that are closely linked to local social structures, this concentration may be unacceptable. For example, in the Atlantic Canada lobster fishery, which has a fixed number of lobster traps per license holder, the fishers strongly oppose trap transferability and more particularly oppose ITQs. These fishers are highly suspicious that the government's promotion of ITQs in other fisheries may intrude into their own successful management structure.

Nontransferability is often seen as a means to maintain the traditional organization of the fleets and fisher communities. Nontransferability may also prevent consolidation of fishing rights in the hands of a few and particularly in the hands of processors or dealers. Maintaining employment of crew may also be an objective. Some increase in efficiency is still possible, because each fisher can catch his or her harvest quota, or utilize his or her gear use allocations, in as efficient a manner as possible. Experience with the enterprise allocation program, under which offshore trawler fleets in Atlantic Canada (primarily owned by two companies) have operated since the mid-1980s, shows that considerable efficiency gains are possible under nontransferable quotas.

Two intermediate options are worth noting. First, not all permits need be transferable. In the lobster fishery of Atlantic Canada, regulations in 1969 defined two classes of licenses: transferable "A" permits for full-time lobster harvesters and nontransferable "B" permits for part-time lobster harvesters. Defining transferability from program initiation meant that policy, not bad luck, determined which fishers ultimately left the fishery without compensation.

Secondly, rather than limit transferability, some management plans limit the maximum amount of rights that can be owned by any person or firm. In some fisheries, such as the Atlantic Canada lobster fishery, this objective of restricting concentration can largely be achieved through an owner-operator requirement (i.e., permit holders must personally fish their own permits). However, such limitations may be difficult to enforce in other cases if the limit can be evaded through ownership by family, relatives, or employees. Contracts that require the nominal owner to follow instructions provided by some third party (such as a processor) can also be used to circumvent the restrictions on ownership concentration. Such restrictive contracts can create a system of dual ownership under which there is a nominal owner for purposes of fisheries regulation and someone else who exercises most of the rights usually associated with ownership. In the scallop fishery of southwest Nova Scotia, the courts have upheld the validity of such restrictive contracts.

To avoid concentration of permit ownership, limitations may be placed upon the ability to mortgage the permit. With the right to mortgage the permit, a fisher wishing to enter the fishery can seek financing through conventional banks. Without that right, the fisher is forced to finance the purchase from his or her own assets. Those with greater financial assets may thus have an advantage in access if conventional financing is not permitted, which would probably be the opposite of the intended effect.

SUMMARY

The choice of the form of user rights depends on many factors including social, cultural, and economic environments (15). Many fisheries from the South Pacific to Nova Scotia have a history of customary marine tenure; with better understanding and nurturing, these may continue to be efficient systems of user rights. Systems of ITQ management have become popular in fisheries where economic efficiency is an important goal, although enforcement, dumping, and high-grading remain serious issues. Gear use rights, which regulate use of some key components of fishing effort, are an attractive means of avoiding the incentives for high-grading and dumping and of reducing enforcement costs. For gear use rights to be effective, a relatively good index of fishing pressure must be defined and periodically adjusted.

Managers and planners must avoid imposing inappropriate user rights systems. Rather they must determine, together with fishers, a form of user rights that will work in practice given the history of the fishery, the attitudes of the fishers, and the nature of the resource.

NOTES

1. Native Council of Nova Scotia (1994).
2. See, for example, Christy (1982) and Ruddle et al. (1992).

3. Acheson (1975).
4. Brownstein and Tremblay (1994).
5. Davis (1984).
6. Ruddle et al. (1992).
7. See Townsend (1990) for a comprehensive review of the experience with limited entry.
8. DFO (1989).
9. But a few economists, such as Copes (1986), are more critical of ITQs.
10. Gardner (1988); Crowley and Palsson (1992).
11. Halliday et al. (1992); see also Chapter 6.
12. Crowley and Palsson (1992).
13. For example, Angel et al. (1994).
14. See Townsend (1992) for one proposal that creates incentives for greater fisher involvement in the appeals process.
15. Charles (1994).

Larry Felt, Barbara Neis, and Bonnie McCay

COMANAGEMENT

Management includes everything from determining the objectives of a specific management plan to participating in the collection and interpretation of the data upon which management decisions are based. Increasing evidence suggests that failure to meaningfully incorporate users is likely to result in high levels of rule avoidance and minimal levels of acceptance by those at whom management is directed. "Comanagement" is a generic term for the various ways in which resource users can meaningfully share management with government. In this chapter we note vulnerabilities in contemporary management regimes, discuss the potential contributions of comanagement to successful fisheries management, and identify some basic requirements for successful comanagement regimes. We present two examples of successful comanagement and address some of its limitations.

CRISIS IN THE FISHERIES

Interest in comanagement in the northwest Atlantic is one response to the recent groundfish crisis. This crisis needs to be placed in context. Marine fisheries are experiencing major biological, economic, political, and social problems throughout the world. Seventy percent of the world's wild commercial marine fish stocks are at the limit of their use or are in trouble (1).

The costs of surveillance and management are escalating relative to industry returns. Conflicts between countries, between regions of countries, between sectors of the fishing industry, and on local fishing grounds are common. Both locally and globally, fish and the wealth they generate are migrating from the poor to the rich. The small-scale and artisanal fishery communities that produce most of the world's food fish are bearing a disproportionate share of the costs of stock decline and economic waste (2).

The multifaceted problems confronting fisheries have prompted a call for innovations in fisheries management (3). The introduction of 370-km exclusive economic zones (EEZs) was supposed to provide the basis for more sustainable, national management of fish stocks. However, some states lack the resources necessary to police their EEZs, and stocks that migrate between political jurisdictions are not fully protected by EEZs. But the problems identified above also characterize many national fisheries that have fully developed and costly management regimes. Advocates of comanagement regimes argue that incorporating resource users into management could help to prevent the spiral of ineffectual management, stock decline, increasing capital costs, and heightened cynicism by harvesters and processors that currently plagues many fisheries.

THE COMANAGEMENT ALTERNATIVE

Three types of reform have been proposed in response to current fisheries crises: an expanded scientific effort and a closer fusion of science and management, as found in such innovative approaches as adaptive management; privatization of fishing rights to let the market do more allocation of fisheries resources and perhaps provide incentives for conservation; and organizational and political changes that bring resource users and dependent communities more directly into the management and science process (4). We speak to the third reform, comanagement, but we believe that the three are neither diametrically opposed nor independent. The effectiveness of fishery management depends on the adequacy of the science behind it, the effects it has on fish and people, and how well its rules are followed and enforced. The general argument for comanagement stresses that knowledge, compliance, and mediation of social and economic impacts are all influenced by how and how much resource users are incorporated into the management process.

Fishery management regimes around the northern Atlantic, as elsewhere, have not only failed to achieve their larger goal of stock recovery and sustainable fisheries, they also have tended to pit managers against the managed. Moreover, "modern" fishery management regimes have often undermined rather than supported longstanding local management regimes associated with inshore and artisanal fisheries (5). There is great variation among nations of the northern Atlantic in how fisheries management is organized, but decision making is usually centralized within government; resource users and dependent communities play little more than advisory roles (6). In the Scotia-Fundy region (northeastern Nova Scotia south to the U.S. border) of Canada's Department of Fisheries and Oceans (DFO), advisory committees of resource users have mushroomed from a handful in the 1970s to dozens covering every sector and exploited species. In the Newfoundland and Gulf of St. Lawrence regions, DFO similarly accords consultative or advisory roles to fishers and their representatives, including unions (7). In advisory systems, user groups are consulted about their concerns and opinions, but decisions are made by the fisheries ministry. This is not comanagement.

Comanagement is not simply more advisory power. In a strict sense it refers to the sharing of power to make decisions, as well as accountability for the consequences of those decisions, with a government agency. Comanagement also presumes a legal framework within which both autonomous and shared decision making takes place. Such a framework is provided in the northwestern United States by the "Boldt decision," which required comanagement of salmonid fishes by Native American tribes and state government agencies (8).

Power sharing need not be simply between large government agencies and smaller, local constituencies. Sometimes central authorities have delegated some or all of their management rights to local political jurisdictions, which then comanage with local interested groups (9). Power sharing is often spread among several levels of government as well as nongovernment constituencies. For example, the management regime for the huge Columbia–Snake river watershed in the Pacific Northwest is based on sharing and cooperation among contiguous states, several federal government agencies, Native American tribes, and the power utilities; in recent years, environmental groups have also assumed a significant role (10).

Management policy imposes both direct and indirect restrictions on individual users in the name of some larger objectives such as resource stability, equity (fairness or social justice), and rent (income to the resource user and manager). Acceptance of management policy depends on the perception that these objectives are being achieved. Conflict frequently exists between short- and long-term objectives, as well as between different sectors of the industry. In addition, scientific uncertainty can make it difficult both to set and achieve stock management goals (11).

Perhaps the most significant benefit of comanagement is the heightened acceptance of and compliance with management rules that can result from meaningful participation in their formulation and revision. This sense of acceptance and compliance is generally referred to as legitimacy. To the extent that rules reflect the experiences and solutions proposed by users and result from dialogue rather than from unilateral imposition by distant bureaucrats, users are less able to rationalize rule violation by referring to "them versus us." Moreover, a growing body of evidence suggests that communities and occupational associations such as fishers have a wide variety of internal mechanisms to ensure compliance with community rules, mechanisms that can be availed of by comanagement regimes. Of course compliance also requires other conditions. The rules must appear to be working. Minimally, this requires that rules are perceived to be preventing stock declines and to be allocating resources fairly. They must also be responsive to industrial change (12).

Some evidence suggests that comanagement can enhance science-based decision making. Proposed reforms of fisheries science to accommodate uncertainty may help reduce some management problems, but science is only as good as the data to which it has access. The size of their jobs and the relative scarcity of their fiscal resources have forced fisheries scientists to rely heavily on catch and effort data from harvesters in their stock assessments. Harvesters, be they corporate entities or self-employed fishers, have contributed to scientific uncertainty and to overharvesting through actions that have distorted scientific data and undermined the effectiveness of management initiatives. Scientific data are distorted by mis- and unreported catches, illicit selling of fish in excess of allocations, high-grading, and illegal use of small mesh sizes and of net liners (13). A general failure to minimize incidental catches of nontargeted species, high-grading, and a tendency of fishers to move to new grounds and switch effort as catches of the dominant commercial species decline are examples of fishing practices that are difficult to monitor scientifically, ecologically dangerous, and poorly regulated within existing management regimes.

Minimal exchange and dialogue between fishers and scientists has meant that important information that helps to explain catches frequently is not considered in stock assessments. For example, changes in fish-locating technology introduced into Canadian offshore fleets in the late 1970s and 1980s allowed Atlantic groundfish trawlers to target fish stocks efficiently and maintain catch levels despite a declining resource—declines that were not immediately recognized because the new fishing efficiencies were not taken into account (14). Such biases limit the ability of science to meet the requirements of politicians and resource managers for consistent, accurate stock assessments.

Scientists are more likely to secure good data and rapid feedback on the ecological effects of management initiatives when resource users are committed to the management process and are active participants within it. Fishers acquire knowledge of local marine ecosystems and the effects of fishing through ongoing experience. Researchers from several disciplinary backgrounds have argued that this knowledge represents an important supplement to scientific understanding and perhaps an alternative foundation for sustainable resource management. To the extent that specific comanagement regimes are grounded in local community management traditions and local knowledge, they can benefit from "rules of thumb" developed from past experience and enforced through established social and cultural means. Such benefits are most clearly seen in long-standing comanagement regimes based on preindustrial arrangements, but they can be realized in more recently established fisheries (15).

Fishers' ecological knowledge could be used in the design of marine sanctuaries for the purposes of habitat and species conservation, tourism, and environmental education (16). More generally, the dialog between fishers and scientists established by comanagement enhances the intuitive understanding of fisheries by both groups, promoting the rapid feedback and social learning that allows flexible, adaptive management strategies (17).

Dissatisfaction with command-and-control approaches to management by government agencies has prompted calls for fisheries reform through both privatization and comanagement. Privatization of fishing rights may bring greater decision-making power to rights-holders, which is an objective of comanagement (18). However, fisher participation will become very narrowly defined if ownership becomes concentrated. It is sometimes argued that privatization will create stewardship incentives for owners of fishing rights. However, privatization may also worsen problems of cheating and noncompliance. With or without privatization, some form of cooperative management may be required to reduce such damaging practices as high-grading, discarding, and informal "under the table" sales.

Comanagement allows fishers to participate directly and meaningfully in management of their resources without necessarily transforming fishing rights into private property. It offers opportunities to address issues that are marginalized in the privatization model: rule acceptance, scientific uncertainty, relations of power, local culture, and social justice.

CONDITIONS FOR COMANAGEMENT

Various comanagement regimes have been and are currently operative throughout the world. A recent extensive review suggests that several hundred cases, at varying spatial scales, currently exist (19). They include the multisector cooperatives in Japanese villages, the cod fishery of Norway's Lofoten Islands, parts of the Newfoundland inshore fisheries, and indigenous management systems in Palau (20).

Like other management regimes, comanagement must include mechanisms for limiting access, determining allocation, resolving conflicting uses, ensuring habitat protection, and ensuring adequate enforcement. Moreover, it must promote legitimacy among resource users as well as compliance and a willingness to exchange information with biologists monitoring the resource. The structure and effectiveness of comanagement regimes are influenced by local political and cultural traditions, the behavioral characteristics of the fish species sought, the range of technology used in harvesting, and the geographic scale of the management unit. An extensive literature on organizational participation and its linkages to perceived effectiveness (efficacy) and acceptance (legitimacy) highlights several conditions that appear to contribute to successful comanagement (21):

- a strong tradition of cooperation and a high level of trust among participants and participant organizations;
- a relatively homogeneous socioeconomic status among participants;
- allocation of some ownership or quasiownership of the resource to participants;
- allocation of substantial decision-making authority to participants;
- equitable allocation of costs or restrictions among participants;
- regulations that make intuitive sense to participants;
- regulation enforcement that involves participants;
- a management regime that encompasses the geographical territory inhabited by key resources and includes all participants whose actions will influence the ecology of the resources; and
- incorporation of fishery workers' ecological and fishery knowledge into stock assessment and fisheries management.

The realization of comanagement regimes based on these and other conditions (22) will depend on the willingness of governments to create appropriate legal and administrative frameworks. We know of no actual comanagement regime in which all of the above conditions are met, but some regimes have met many conditions and have been studied in sufficient detail to outline their main features. Two of the best known are the Japanese cooperative associations and Alaska's cost recovery salmon enhancement associations.

CASE STUDIES OF COMANAGEMENT

Japanese Fisheries Cooperative Associations

Most comanagement regimes throughout the world are limited in geographical size and number of membership groups. However, a major portion of Japan's national inshore and nearshore fisheries is comanaged through Fisheries Cooperative Associations (FCAs [23]). These fisheries are dominated by small-scale and household enterprises. In 1989, they contributed one-third of the country's annual catch and 50% of its value. The FCA system of community cooperatives grants exclusive legal rights, under national legislation, to the harvest of designated resources in a defined territory. In 1990, approximately 2,127 FCAs had a combined membership of 535,000 people. The average FCA has about 250 members. Local FCAs are linked to larger regional and national associations.

Japanese inshore fishery management is derived from a nearly 1,000-year-old tradition of decentralized political authority and strong local occupational associations. Three laws, based on four major principles, provide the basis for the entire comanagement regime. These principles are exclusive-use rights in nearshore zones, control of fishing intensity of various fisheries in defined waters during specified seasons, establishment and enforcement of various conservation measures, and promotion of coordination and cooperation among fishers.

The FCAs are legally entrusted with exclusive territorial rights for management of sedentary resources such as mollusks, seaweeds, and local fish stocks. Fishers, in cooperation with a few government agencies, have the responsibility to define management rules and their implementation and enforcement. Associations also develop management plans for each fishing right they hold, including aquaculture activities. Once developed and approved by the membership, these plans are then reviewed and ratified by a local Prefectural Fisheries Agency and the nearest of 65 regional Sea Area Regulatory Commissions. The majority of seats at both levels are filled with representatives elected by FCAs; the remainder are appointed by prefectural governors.

More than 2,000 FCAs each hold multiple rights to different marine resources and technologies, so up to several thousand specific territorial rights may be held by FCAs along a 50-km section of the Japanese coast. This level of complexity precludes simple top-down management. Broader management organizations deal primarily with dispute settlement. The United Sea Area Fisheries Adjustment Commissions, for example, are typically created to settle jurisdictional disputes.

Although FCAs inevitably differ, they seem to be comanagement regimes that have led to sustainable fisheries and high levels of income for fishers. This regime is able to address the management problems associated with successional fishing, irregular changes in stocks, bycatch, and harmonious and systematic uses of whole ecosystems, and it has become an important component in general regional economic development.

Alaskan Cost Recovery Salmon Enhancement Associations

In contrast to the inshore, relatively sedentary resources managed in the Japanese system, many marine fisheries target highly migratory species such as salmon. Salmon may migrate thousands of kilometers from their rivers of origin before returning to those rivers to reproduce. Commercial harvesting occurs throughout the migration route, on the high seas, in coastal zones, and within the rivers themselves. Salmon management is further complicated by different harvesting technologies, the need to allow sufficient fish to escape the fisheries for reproduction, and a requirement to allocate harvests among (usually) aboriginal, commercial, and recreational sectors. Despite the complexities of salmon management, Alaskan Cost Recovery Salmon Enhancement Associations have exemplified relatively successful comanagement (24).

Alaska has a coastline of approximately 45,000 km—more than the remainder of the United States shorelines combined. At the time of her admission as the 50th state in 1959, harvests of her once-abundant salmon runs had dropped from an annual average of 140 million fish to a record low level of 30 million harvested fish. In the 1960s and 1970s, the state legislature began several programs to rebuild stocks and to provide for their sustainable management. One initiative was legislation enabling the creation of regional enhancement associations made up of commercial fishers.

The enhancement associations were allowed to borrow money from the state to undertake programs and to sell surplus salmon from these projects to pay for their activities. The associations also had the authority to become partners with the state in developing long-term enhancement plans. In the past 20 years, at least five associations have successfully increased the number of salmon available to local fishers and hence increased member incomes. In addition, the regional associations have assumed a greater informal role in regional strategic planning, stock assessment, allocation, improved product quality, and alternative marketing. One regional association plans to build a research laboratory in which such tasks as otolith analysis (for aging fish and distinguishing wild from hatchery-reared fish) will be done. Over the years, membership in this comanagement regime has been extended to include representation from other groups. Commercial fishers still constitute a majority of participants, but Native Americans, recreational fishers, processing plant owners, and even the general community are also formally represented.

LIMITATIONS OF COMANAGEMENT

Comanagement offers the potential to improve management regimes for many species and in numerous jurisdictions, but it is not a magical panacea for the worldwide fisheries crisis. Comanagement works best where stocks are sedentary or migrate modest distances and where traditions of local control and cooperation exist. It may be possible to implement a division of labor in which local users and their associations participate directly in local management issues (allocation, local assessment, etc.) and simultaneously have meaningful participation in resolution of broader issues via larger regional, national, and international bodies.

The communal management regimes that preceded massive state intervention into fisheries management were generally associated with local, relatively low-technology fisheries. Comanagement regimes for northwest Atlantic groundfish would need to address the difficulties associated with managing heavily depleted migratory stocks and address the technological, social, and institutional legacies (including the scientific legacy) of relatively open-access, nonselective, mobile fisheries.

A movement toward more participatory fisheries management can be discerned in the northwest Atlantic. Although imperfect, it is evident in the U.S. regional fishery management council system. The councils, composed of both government officials and members of the public representing various fisheries interests, have responsibility to develop management plans for fisheries in the EEZ.

They do this in a very open, public process. A federal agency has the ultimate right to approve or disapprove the plans and the responsibility to implement them. The new Canadian Fisheries Resource Conservation Council (FRCC) also embodies more participatory management. Government-appointed members of industry, the scientific community, and government consult with stakeholders and make public recommendations to the fisheries ministry. Like U.S. fishery councils, however, the FRCC is controversial. Strong industry participation can lead to allegations of "foxes in the henhouse" (25). Appointed councils raise concerns about the adequacy of representation: Who selects the individuals who will serve? Are all legitimate constituencies represented? In Canada, both participant selection and the definition of "stakeholders" are controlled by government, and no explicit criteria for either have been established.

Most writers on fisheries comanagement appear to assume that comanagers should be limited to government, industry, and harvester representatives. For the northwest Atlantic, several industry groups and some government agencies have agreed that comanagement of this type is necessary, at least in Canada (26). Nevertheless, an argument can be made for involving other groups. Decision-making structures need to be relatively efficient and effective and decision-making processes need to be clearly defined. However, new comanagement regimes also need to address social and economic inequalities that are products of past regimes. Examples are the frequent marginalization of women, young people, the elderly, and crew members within existing decision-making structures. Older community residents can provide ecological knowledge and memories of the potential abundance of currently depleted resources. Women, because of their responsibility for managing fishery households and economic vulnerability, should be involved in resource management; they can bring unique insights to the table. Similarly, young people, because of their dependence on the future of the resource, and crew members, because of their vulnerabilities to fishery wealth, could all add important insights and commitments to an effective comanagement regime. These groups often played a greater role in past community-based management systems than they do under recent state regimes (27).

SUMMARY

A palpable conclusion from present fisheries crises around the world is that sustainable fisheries management requires full cooperation and support from those who are being managed. Cooperation is essential for effective compliance with regulations as well as for maximizing the quality and range of knowledge upon which management is based. Comanagement is the best way to secure such cooperation. Comanagement is, however, a general management orientation that must be customized for specific industrial, political, environmental, and cultural situations, as well as for particular behavioral characteristics of targeted marine resources. Meaningful comanagement requires transforming established institutions and, as a consequence, reallocating power. There is a significant risk that those who dominate current management regimes will co-opt the comanagement concept for their own ends, dressing up current practices of scouring the shoreline for advice as a substitute for devolving power. Such "comanagement" regimes could squander whatever remains of our wild fisheries and further alienate those who are managed. Although it offers no immediate or simple solution to contemporary management failures, real comanagement addresses a wider range of management problems than any other general strategy and it has proven to be effective in parts of the world beyond northwest Atlantic fisheries. Where feasible, comanagement should be pursued as the foundation of future management systems (28).

NOTES

1. Global fisheries statistics are compiled by the Food and Agriculture Organization of the United Nations (FAO 1994). Nielsen and Wespestad (1994) described the current crisis as follows: "Exploitation, habitat degradation, and inappropriate management have brought dozens of species and scores of stocks to such low abundance that their subsistence, commercial, recreational, and ecological values have dripped away like water leaking through cupped hands."
2. Kent (1987); McGoodwin (1990). These costs include displacement of thousands of workers (environmental refugees) and, in cases like the northwest Atlantic fisheries crisis, erosion of the socioeconomic basis of entire regions.
3. Ludwig et al. (1993).
4. Walters (1986), Ludwig et al. (1993), and Sissenwine and Rosenberg (1993) are among those who have urged closer integration of science and management; see also Chapter 7 of this volume. Neher et al. (1988) and Mace (1993) have addressed privatization issues, which are given additional treatment in Chapters 10 and 12. Jentoft (1989), Pinkerton (1989), and McCay (1996b) have described features of comanagement, which is also evoked in Chapter 5.
5. Johannes (1978, 1981); Matthews (1993).
6. Jentoft and McCay (1995).
7. MacInnes and Davis (1996).
8. The Boldt decision, named for the presiding U.S. federal judge in a case brought by tribes to reestablish treaty rights to salmon, has been discussed by Cohen (1986) and Pinkerton (1989, 1995), among others.
9. Jentoft and McCay (1995).
10. Lee (1989).
11. Holling (1978); Walters (1986).
12. Various aspects of comanagement legitimacy have been discussed by Kearney (1985), Dyer and McGoodwin (1994), Pinkerton (1994), Pinkerton and Weinstein (1995), and MacInnes and Davis (1996).
13. Distortion of stock assessment data by fisher practices was discussed by Alverson et al. (1994). See also Chapter 3.
14. This example is from Finlayson (1994).
15. The use of fisher knowledge in scientific stock assessments has been discussed, promoted, or both by Johannes (1981, 1993), Berkes (1987, 1988, 1993), Freeman (1989a, 1989b), Wilson et al. (1990), Gadgil and Berkes (1991), Kloppenberg (1991), Wilson and Kleban (1991), Johnson (1992), Gadgil et al. (1993), Mailhot (1993), Felt (1994), and Walters et al. (1995).

16. Gadgill et al. (1993); see also Chapter 7.
17. Holling (1978); Freeman (1989b); Pinkerton (1994); Maguire et al. (1995).
18. Scott (1993).
19. Pinkerton (1994).
20. Johannes (1978, 1981); Jentoft (1981); Nagasaki and Chikuni (1989); Matthews (1993); Pinkerton (1994); Pinkerton and Weinstein (1995).
21. Jentolf (1989); Felt (1990); Doulman (1993); Pinkerton (1994).
22. Ostrom (1990); McKean (1992); Pinkerton (1994); Pomeroy and Williams (1994).
23. The Japanese FCA program has been described and analyzed by Nagaski and Chikuni (1989), Morisawa and Yamamoto (1992), Ruddle (1994), and Pinkerton and Weinstein (1995).
24. Pinkerton and Weinstein (1994).
25. McCay (1996a).
26. A movement in the direction of broader-based comanagement has taken hold in Nova Scotia via the Coastal Communities Network (CCN; Post Office Box 539, Waverly, Nova Scotia B0N 2S0), which is assisted by the Extension Department of St. Francis Xavier University. At a 1994 workshop convened by CCN, the Scotia-Fundy regional DFO director spoke of that agency's move toward more comanagement. The 1994 and subsequent 1995 CCN workshops explored the options for community-based comanagement in Nova Scotia. Among identified concerns were the narrow federal definitions of comanagement, which seemed to exclude vessel crews, fish processors, families, community economic development groups, and others. The CCN effort is linked to those of other groups, including the new Atlantic Women's Fishnet (40 Scott's Road, McGrath's Cove, Nova Scotia B0J 3L0). Meanwhile, "round table" discussions in Montreal, Quebec, and Bedford, Nova Scotia, during March 1995 identified industry representation in fisheries management and the power to determine representation (and meeting agendas) as two major issues in the Canadian comanagement debate.
27. Doulman (1993).
28. We have been influenced by unpublished ideas presented at the 1994 (Halifax, Nova Scotia) and 1995 (Tampa, Florida) meetings of the American Fisheries Society by C. Bailey and A. Davis; R. Haedrich, D. Schneider, and J. Hutchings; and C. T. Palmer and P. R. Sinclair.

Ralph E. Townsend

CORPORATE MANAGEMENT OF FISHERIES

Open access has been the historic condition of most fisheries. The negative effects of open access upon both resource conditions and the economic condition of fisheries are well documented. State control of the resource has become the major alternative to open access in developed countries in the Twentieth Century. But the record of state control of fisheries has been modest at best, and the search for new approaches to fisheries management has shifted to concepts of private and communal property.

Most alternatives to state control of fisheries seek to involve the fishers themselves more directly in management. The reasons for preferring active fisher involvement are easy to identify. Fishers may be able to devise and to administer regulatory institutions that are superior to externally imposed regulations. The local industry has extensive information about the resource and the harvesting technology that may be very useful in designing effective rules. Once a set of rules is in place, the regulatory authority must enforce those rules. Locally imposed rules may have the advantage of greater local acceptance, and hence may face less opposition from fishers.

To some, the obvious forms of collective self-governance are cooperative management or the comanagement approach of shared governance between local communities and formal government agencies (1). But collective governance can also be achieved through the private institution of corporate governance. The corporation is a collection of resource owners who must make unified decisions about jointly owned assets. The corporation is the dominant institutional arrangement for economic activity in most of the developed world. Given the dominant role of corporations in our economies, corporate governance of fisheries certainly deserves consideration.

CORPORATE GOVERNANCE OF FISHERIES

To establish a corporate structure for governing fisheries, the government would first identify the resources for which the right to fish was to be established. The government would then set the rules that determine who will become the initial owners of the fisheries governance corporation. These corporate shares could be given to some set of fishers who already hold licenses or individual transferable quotas (ITQs) or they could be sold at auction to generate a return to the public for use of the resource.

For example, in a fishery currently managed via 150 limited-entry permits, a corporation with 15,000 shares would be created if each existing permit holder were given 100 shares. Multiple shares should be created for each owner because divisibility of interests is frequently necessary. If two share owners want to buy the interest of a third owner, the ability to divide an interest facilitates this transfer. If a business partnership or a marriage dissolves, a court-ordered division of assets would be facilitated by easily divisible rights.

The bylaws of a fishery management corporation would contain three constraints on all decisions. First, fishing benefits must be bestowed in proportion to share ownership. For example, if the corporation implemented ITQs, quota allocations must be in proportion to share ownership. If fishing effort were controlled, the allocation of fishing effort must be in proportion to share ownership. Second, fees imposed to finance the corporation must be levied either in proportion to ownership shares or in proportion to landings. These first two provisions would protect minority shareowners from discriminatory actions by a majority (2). Third, the board and the staff of the corporation would be subject to greater requirements for auditing and public reporting than typical corporations. The fishery management corporation would have to provide the government with financial and biological data, audited when necessary, to insure that the terms of the operating contract (described below) have been met.

The fisheries governance corporation would decide all aspects of management, including how to regulate fishing, who would be allowed to fish, and how to implement and enforce decisions. If the number of owners is small, decisions about management could be made by direct voting of shares. For most fisheries, however, the election of a board of directors with the authority to govern the corporation would be more practical. Through corporate governance, the owners would decide what management approach—such as effort controls or ITQs—would be appropriate. The corporation would decide when to invest in stock growth (by delaying harvest) and when to harvest. Owners of shares in the corporation would be free to sell their rights or to lease to third parties the fishing rights attached to their shares.

Fishing activity under corporate management could be conducted in at least three different ways. The corporation could authorize and regulate fishing activity by its owners. Under this approach, the collective decision making of the corporate owners would be analogous to current government regulation. This is probably the model most corporations would adopt. Alternatively, the corporation might decide to lease fishing rights through a competitive auction, the revenues being distributed to owners in proportion to share ownership. Allowing nonowners as well as owners to bid in such auctions would maximize the economic return to the corporation owners. Finally, in some fisheries, the corporation might act as the vessel owner and operator on behalf of all members. The total profits of the corporate fishing enterprise would be shared with owners in proportion to ownership. Such vertically integrated fishing activity might be appropriate in industrial fisheries where large economies of scale exist in harvesting.

A fisheries governance corporation could be expected to finance its own operations, including administration, enforcement, and biological stock assessment. Industry financing of these costs is not typical of fisheries today, and industry opposition to assuming this responsibility can be expected. These management corporations would create the opportunity for fishers to realize significant profits, from which administrative costs could be paid. By internalizing the costs of implementing and enforcing management, the corporation has incentives to minimize these costs when making management choices. Moreover, greater fisher responsibility and the associated opportunities for greater income will be politically more acceptable if there is concurrent responsibility for the costs of administering fisheries.

The corporate structure with transferable share rights creates a well-defined, long-run interest in the resource. The ability to transfer well-defined corporate rights facilitates investment decisions (such as whether or not to invest in stock rebuilding). Those who are willing to make such an investment can buy shares from those who do not favor the investment. In so doing, these investors not only incur all the risk of the investment, they also compensate the previous owners for the opportunity to make the investment. In return, they receive all of the future benefits. Also, because rights are clearly defined, stockholders will be able to use the corporate shares as collateral for debt financing.

The market for the shares of the corporation creates an important bias in favor of conservation. A fish stock is an asset that provides a stream of harvests (and hence income) into the future. There are trade-offs between current harvests and future harvests that must be assessed, and different economic actors will have different perspectives on these trade-offs. Investors who place the highest value on the future will be willing to pay more for the long-term stream of harvests (and income) than those with more myopic planning horizons (3). An owner who wants immediate returns will earn more by selling his or her rights to someone with more interest in future income than by overfishing the stock. Hence, market transactions will result in ownership of these assets by those with the greatest concern for the future. If fishers have shorter planning horizons than investors more generally, then rights will tend to be purchased by nonfishers. Note that the ability to mortgage the fishing rights will facilitate the financing of purchases by investors. For these reasons, ownership of shares in fishing corporations should not be restricted to fishers nor should the ability to mortgage the asset be limited.

It might seem that there is little practical difference between corporate management and collective management or comanagement. These institutions may share some features, but they differ fundamentally in governance structure. Corporations are governed by one-share–one-vote principles. Shares are freely transferable, which creates a market process for decision making. All costs and benefits of corporate ownership are strictly in proportion to share holdings. In contrast, collectives and similar institutions are governed by democratic processes (one person, one vote). The inherent uncertainties of the political process create uncertainties about future distributions of costs and benefits to the collective members.

To understand the differences between a cooperative and a corporation, consider a decision that would close a fishery in order to achieve higher catches in the future. Under the democratic voting of a cooperative, a majority of the current members must be convinced to incur the immediate costs of closure. But the sharing of the ultimate benefits will be determined largely by votes taken after the resource has recovered. Risk-averse voters will be reluctant to incur these costs in exchange for benefits whose future distribution is uncertain. Obviously, the more open the membership structure of a cooperative, the more serious this problem becomes. The problem is likely to be compounded when participants are economically or socially heterogeneous or geographically dispersed. But even if membership is strictly closed, members who must incur greater costs now (e.g., those with larger boats or greater financial dependence upon the fishery) have no assurance that they will be rewarded with greater benefits later. This problem of short planning horizons for democratically governed institutions has been extensively analyzed for both agricultural cooperatives and worker-owned firms (4).

Inasmuch as many fisheries face the task of rebuilding stocks, the democratic management of cooperatives seems especially unsuited to the task at hand. In these crucial decisions, the differences in the time horizons of corporate owners versus cooperative members are likely to be clearest. Because the corporate shares could be used as collateral for loans, corporation owners could also borrow more easily. The ability to borrow facilitates decisions not to fish for some period while stocks rebuild.

PRESERVING THE PUBLIC INTEREST

The objective of creating efficient incentive structures through corporate property rights is easily stated, but the task of defining precisely what rights will be bestowed is not trivial. Typically, the right will be to use a particular species or set of species for commercial harvests. Because these species live in a broader ecosystem, the decisions made by these property owners will affect other elements of the system. Some of these interactions are as direct as bycatches of endangered species; some are as diffuse as the social interest in biodiversity. The initial definition of rights must specify the responsibilities of the fisheries management corporation in these areas, as well as its rights to fish.

Because the knowledge of ecological interactions is limited, it is inevitable that the initial allocation of rights will not address all of the interactions created by the fishery resource owners' actions. The initial property rights system must address the question of not only what rights are bestowed, but also how definition of those rights may evolve in response to changing circumstances or information.

A tempting, but seriously flawed, approach to protecting the public interest would be to subject the decisions of a management corporation to routine approval by some government agency. A government veto has a corrosive effect upon incentives and will almost certainly lead back to the status quo of government command-and-control regulation. Continuous government oversight will inevitably become continuous government regulation.

A more effective way to create a property right with flexible dynamic characteristics is to implement a fixed-term operating contract between the fisheries management corporation and the government. Under this institutional arrangement, the rights given to share owners would be permanent, but certain conditions on use of the resource would be subject to periodic renegotiation and revision. The framework of renegotiation must provide the maximum incentive for the owners and the government to negotiate successfully. To face truly long-run incentives, the owners require "smooth" and predictable outcomes from this negotiation. For this objective to be met, the government cannot have the preponderant power over contract terms.

The renegotiation framework must create rights and expectations that are well defined. Some fairly broad rules are clear: the government cannot replace the owners with new owners, nor can the government impose any fee not specified in the initial property rights definition. The government retains the responsibility for certain broad public interests, such as biodiversity, ecosystem health, and species protection. This responsibility is two-sided, in that the government also has the responsibility to regulate activities that negatively affect the resource. The government's public interest rights might be subject to some requirement for a scientific basis for any restrictions. The framework should also specify the extent to which compensation is due for changes in the operating contract. If, for example, the industry buys the initial rights at auction, the governing framework might specify that compensation will be paid if rights are restricted. These principles still leave broad gray areas, but they substantially restrict the areas for potential disagreement.

The operating contract would not remove the obligation of the fishing industry to obey general health and safety regulations of the government. For example, changes in boat safety regulations would still apply to the fleet during the period of the contract. If all seafood from some area were embargoed because of environmental contamination, such as an oil spill, the industry would be subject to that embargo.

An initial fixed-length contract of 10 to 20 years would be specified. The exact length can vary, but it should be sufficiently long that investments in both physical capital and resource status can yield adequate returns. A reasonable rule of thumb might be that the contract length should exceed the greater of the technological life of fishing vessels or two reproductive generations of the biological stock.

But any fixed-length contract can distort incentives as the end of the contract is approached. In fisheries, the obvious problem would be incentives to overfish the resource as the terminal date of the contract approached. Neither the government nor the resource owner wants this incentive to become dominant. For that reason, the two parties would enter into negotiations for renewal of the contract at the midpoint of the contract (e.g., at the beginning of year 10 in a 20-year contract). The new contract would become effective at the time of renegotiation (e.g., at year 10). If contract renegotiation fails at this midpoint, both parties have the option of continuing under the original contract until the expiration date (e.g., year 20). Under this system of staggered contract renewal, strong incentives are created for both parties to reach agreement (5). Both parties gain if a contract is renewed at the earlier date. The government obtains the benefits of any new environmental or resource protections added to the contract at year 10, instead of waiting to impose them at year 20. The owner gets the security of a new 20-year planning horizon. Both parties have considerable leverage in the bargaining process. The government can implement new (and perhaps more onerous) requirements at the end of the contract if renewal negotiations fail. The owner can simply allow the existing contract to run to its end, thus thwarting additional restrictions for the remainder of the contract. In a sense, both parties can use time as leverage in the negotiations.

The operating contract accommodates dynamic issues in the definition of property rights. The owners' rights can evolve in response to our understanding of the resource and its relation to other resources, just as other rights evolve. Within the term of the contract, and given the restrictions on its redefinition, the owner can make investment decisions under a relatively long and secure planning horizon.

A PHILOSOPHICAL CHANGE

Fisheries require government regulation because open access creates economic incentives that result in economic and biological overexploitation of the resource (6). The problem is not the greed of individual fishers or any unique technical characteristics of fishery stocks; the problem is the inappropriate incentive structure under open access. Government has historically tried to correct these incentives by command-and-control regulation. Corporate management, by contrast, rearranges the institutional framework so that the resource owners have appropriate incentives for wise resource use. Corporate management matches the solution to the underlying institutional problem.

Because corporate management creates incentives to manage the resource appropriately, owners can be given a broad range of decisions that are currently reserved for government. These include determination of total allowable catches, of gear or mesh restrictions, of area or seasonal closures, of effort limitations, and of resource allocation among users.

As the negotiator and enforcer of the operating contract, government would have a new and more limited, but still important, role. That contract regulates the negative effects of fishing activity upon resources not owned by the management corporation. For example, interactions with endangered or protected species and with other fishery resources would be governed by the operating contract. This is analogous to the government role in the regulation of other private resources, such as forests. The government regulates forestry practices to prevent damage to streambeds and the watershed, to protect endangered species and biodiversity, to manage game species, and to regulate the negative aesthetic effects of logging practices. The government does not regulate forestry practices in order to dictate strictly economic decisions to the private landowner. The individual landowner has incentives to maximize the economic return from his or her forest. Government regulation is directed at the noneconomic returns to the rest of society.

For those who are comfortable with government regulation of fisheries, this change may seem especially radical. The risk that fishers as owners will make "mistakes" may seem a risk that can be avoided by imposing government decisions. But regulation has its own very real limitations that must be factored into the choice of management approach. The relevant institutional comparison is not between theoretically perfect regulation and imperfect decision making by real-world private owners, or even between theoretically perfect private decision making and theoretically perfect public regulation. The relevant comparison is between actual performance of imperfect humans under government regulation and performance under private corporate management. To make that comparison, fisheries managers must be prepared to experiment with corporate management. The greatest opportunity for success (and the lowest cost of failure) for corporate management will be in fisheries where a small, cohesive group of fishers has already taken an active role in shaping public regulation. Such fisheries are obvious candidates for testing and developing the concept of corporate management.

SUMMARY

This paper has analyzed a new approach to fisheries management: a corporately organized and privately controlled management structure, which operates within long-term contractual requirements for environmental protection. The corporate management approach invokes the same set of private incentives that a market economy relies upon for management of most of its economic resources. This corporate approach shifts the role of government from command-and-control regulator to architect of appropriate social institutions (7).

NOTES

1. See Jentoft (1989) on cooperative management and Pinkerton (1989) and Chapter 11 on comanagement.
2. Protection of minority rights is an established principle in corporate law. Using an established institution like the corporation has the great advantage of incorporating an existing body of law that already addresses many potential problems.
3. Economists use the idea of a "discount rate" to explain decision making about future streams of income. A discount rate is the rate of interest that an investor demands on investments. That rate of interest determines the value of a stream of income. For example, consider a fishing right that can yield $10,000 per year in perpetuity. If an investor expects a 10% rate of return, that stream of income is worth $100,000. To verify this, note that the rate of interest (10%) times the asset value ($100,000) equals the annual return ($10,000). If a different investor expects only a 5% rate of return, then the $10,000 per year has a value of $200,000. Again, the rate of interest (5%) times the asset value ($200,000) equals the annual return ($10,000). A lower interest rate means that the investor places a higher value on returns in the future. As this simple example illustrates, investors with lower interest rates (and who care more about the future) place a higher value on any given stream of income and hence can bid more for ownership rights to that asset. Note also that an owner with a high discount rate can earn more by selling the undepleted resource to someone with a low discount rate than by overharvesting. If, in the simple comparison above, the high-discount rate owner could earn $30,000 in the short term by seriously overharvesting, that would still be less than the gain in the value that could be realized by selling the resource. That gain would be $100,000, the difference between the $200,000 value to the low-discount rate owner and the $100,000 value to the high-discount rate owner.
4. See reviews by Staatz (1989) and Bonin et al. (1993) on agricultural cooperatives and worker cooperatives, respectively.

5. The idea of an overlapping contract renewal process is one that M. Young, of the Commonwealth Scientific and Industrial Research Organization in Australia, has suggested in a variety of public lease situations. See Young (1992).
6. Gordon (1954).
7. This research was supported in part by the National Sea Grant Program under grant NA36RG0110-01 through the Maine–New Hampshire Sea Grant Program, project R/FMD-237, and in part by the research program in Property Rights and the Performance of Natural Resource Systems of the Beijer International Institute of Ecological Economics, the Royal Swedish Academy of Sciences, Stockholm, with support from the World Environment and Resources Program of the John D. and Catherine T. MacArthur Foundation and the World Bank. I acknowledge the very important contributions that Sam Pooley has made to the development of my ideas on corporate management.

Daniel E. Lane and Robert L. Stephenson

13

DECISION ANALYSIS

Successful groundfish management by whatever means will require difficult decisions. Fisheries management problems are complex, multidisciplinary, and wide-ranging in scope. Management decisions must be made at all levels from the fishing vessel (What gear is best suited to catching the fish? How much fish should vessels be allocated?) to the entire fishery (What is the maximum amount of fish that may be caught in a given season? How should this total be allocated among competing harvesting gears?). Difficulties in effective decision making in marine fisheries arise from the inherent variability of oceanic ecosystems, the invisibility of fish stock dynamics, multiple and conflicting objectives, and bureaucratic frameworks that are not responsive to required rapid change (1).

In order to make appropriate decisions in fisheries, managers need a means of evaluating what constitutes "good" and "poor" decisions. "Decision analysis," the formal area of study of complex problems, provides a methodology for such evaluations. Decision analysis has emerged as an important area of social science. It has modified the ways in which conflicts are resolved in such areas as industrial labor–management disputes, trade bargaining, and arms negotiations, and in which decisions are made on such major projects as oil drilling, airport planning, and financial analysis (2). Despite the well-known difficulties of fisheries decision making, few examples of structured approaches to problem solving exist in fisheries management. Institutional fisheries arrangements, which tend to be constructed along disciplinary lines, have been unable to adapt to the increasingly interrelated and multidisciplinary issues of modern fishery problems. Efforts to deal with evolving problems have typically focused on discipline-based analyses, and the resulting patchworks of regulations and support programs have not addressed fundamental aspects of the problems. In this chapter, we describe a general framework for decision analysis in fisheries management, develop the process for effective implementation of such analyses, and describe appropriate supporting institutional arrangements.

FISHERIES DECISION ANALYSIS

In Clemen's words, "The purpose of decision analysis is to help a decision maker think systematically about complex problems and to improve the quality of the resulting decisions" (3). Fisheries problems are complex for several reasons. They comprise many different but interconnected elements such as fish biology, ecosystem effects, commercial fishery dynamics, and social, economic, and political forces. They embody several, often conflicting objectives such as stock conservation, commercial exploitation, and employment. And they are complicated by environmental shifts, market fluctuations and substitutions, weak year-classes of fish, and other large and small events that are unpredictable and uncontrollable by management.

A decision analysis framework treats complex problems in a structured way by decomposing them into understandable elements: (i) definition of the problem in clear statements, (ii) generation of alternatives for problem resolution, (iii) evaluation of the potential effectiveness of alternative decisions, and (iv) monitoring of the actual effects of the implemented decisions in relation to expected effects.

Problem definition requires a clear statement of the problem along with a concise definition of the form of the decision needed to resolve the problem. All problems are characterized by sets of "variables" that are either controllable (through an appropriate decision selection) or uncontrollable (unaffected by decisions). Comparison of key problem variables (e.g., spawning stock biomass, stock–recruitment relationships, fishing revenue and costs, employment levels in the fish processing sector) permit decision alternatives to be evaluated with respect to desirable objectives.

Clear enunciation of objectives is required in order to measure and compare the effectiveness of decision alternatives. Objectives are measurable values that describe benefits derived from, for example, the social and economic activities of the fishery sector (e.g., profits from fishing). Definition and enumeration of constraints, such as a minimum spawning stock biomass, are also required elements of problem definition.

The suite of problem elements—decision definition, constraining factors, controllable and uncontrollable variables, and measurable objectives—complete the required input to the decision analysis process. However, problem information is not static, but dynamic. As the decision process evolves, more information about the problem is compiled and included in the analysis.

The activities of the decision analysis process include (i) compiling problem information, (ii) analyzing the pertinent data, (iii) developing decision alternatives, and (iv) evaluating each alternative against the stated problem objectives. The result of this process is a suite of decision alternatives that can be presented to the decision maker in support of the decision to be made.

The decision process does not end when the decision is made. The implementation and impact of the decision must be measured over time. Monitoring and appropriate adjustments in implementation should be consistent with the measurable objectives established during problem definition.

Defining the Problem

Fisheries management problems require decision making on two levels. First are the "global" problems. In the case of groundfish management, these relate to regulators' responsibilities for determining annual total allowable catches (TACs) for individual groundfish stocks (4). On the second level are operational and seasonal management decisions (e.g., the timing and location of area closures) required to ensure that the regulators' designated TAC limits are not exceeded and that the ongoing issues are resolved efficiently. Examples of these decisions are vessel or gear suballocations, enforcement and catch monitoring, and licensing. The fishing industry is directly implicated in these problems and likely will acquire more responsibility for these operational aspects of the fishery. Recent discussions about cooperative management or "comanagement" schemes and government–industry partnerships in the Atlantic fisheries could lead to much more responsibility and decision-making power for the industry than exists under current advisory and consultative management structures (5).

By way of example, a simplified version of a TAC-setting problem may be stated as: "At what levels can annual exploitation be set over a planning period such that (i) stock abundance is not adversely affected and (ii) levels of commercial exploitation can occur for the benefit of the fishery sector and the public as a whole?" This problem definition contains all of the major elements of the decision problem. "Decisions" in this case would take the form of a singular global control variable for a harvesting strategy, denoted by a schedule of annual TACs options over the predefined planning period of (for example) 5 years. The length of the planning period is important; it should be long enough for the fishery to evolve toward a desirable longer-run position (including adjustment periods for stock status and industry performance). Controllable elements include variables that may be manipulated directly or indirectly (e.g., the level of the TACs, gear sector allocations). Uncontrollable problem elements include, for example, all environmental factors that contribute to the inherent uncertainty of natural systems.

Biological considerations in fisheries management may be appropriately treated as "hard" constraining factors. Constraining conditions, such as a minimum spawning stock biomass, must not be violated by any feasible strategy. A minimum spawning stock size is an example of a constraining factor that may be necessary for conserving future fishery productivity. Such constraints eliminate decision alternatives that are judged likely to violate the specified conditions. As Wooster stated,

> To understand the process, we first examined the objectives of management, especially the biological objectives, which we thought could be defined unequivocally, whereas social welfare objectives, being heavily loaded with values, would be more controversial. But it soon became apparent that biology imposed *constraints* rather than inspiring *objectives*. Fisheries were managed to obtain social, not biological, benefits, although the magnitude of the benefits, both now and in the future, was constrained by the continuing productivity of the resource. (6)

Stock target constraints might take the form of annual schedules of desirable absolute stock levels for major stock subgroups such as juveniles and adult spawners. Longer-term strategic planning requires specification of short- and intermediate-term outcomes that are consistent with the longer-term goals. For example, in the Atlantic cod fisheries off Labrador and Newfoundland, specific stock targets were set for fishable biomass (fish of age 3 and older) and spawning stock biomass (age 7 and older) out to the year 2000 (7). Recent changes in the U.S. Magnuson (now Magnuson-Stevens) Fishery Conservation and Management Act specify absolute minimum levels for spawning stock biomass as conditions for commercial exploitation. Constraints of the decision problem separate possible decision alternatives into feasible and infeasible categories. The primacy of biological constraints relegates contravening strategies to the infeasible category without further evaluation.

Setting Objectives

Objectives are determined by considering benefits derived from fishing. As Larkin put it,

> The approach must be anthropocentric. It is a contradiction in terms to speak of biological objectives of fisheries management. Much more logical is to speak of biological constraints to management.... The real questions are: what should be the biological constraints and what should be the social objectives. The answers are: whatever is necessary to preserve future biological options until we know more biology and, whatever seems appropriate to the society at the time. (8)

Objectives of the fisheries management problem are measurable valuations to be applied to feasible decision alternatives. These measures are used to compare the expected performance of alternative decisions. Fisheries management problems typically have several and often conflicting objectives that include the following.

- Economic objectives address the performance of commercial fishery sectors. An objective might be an equitable distribution of exploitation benefits among users such that average profitability is consistent among sectors.
- Social objectives address the sustainable public benefits derived from ownership and consumption of the resource. An objective might be to maintain target levels of seasonal employment in fish harvesting and processing.
- Biological objectives address the attributes of a stock beyond minimum abundance constraints. An objective might be a desirable age composition.

Developing and Evaluating Alternatives

The process of developing and evaluating decision alternatives is the end result of the decision analysis process. To achieve this end, a mathematical model of the fishery system is required to project and estimate future impacts of current decisions, and the model must link economic with social and biological objectives. Furthermore, in order to assure model integrity, efforts must be made to incorporate all available problem data into the evaluation process. For example, aggregate "top-down" estimation of biological stock size by biologists should be combined with extensive "bottom-up" evaluations of stock abundance over space and time received from active fishers throughout the season. Such a combination is presently uncommon; for example, TACs for Canadian Atlantic groundfish are determined primarily by a top-down analysis that leads to harvest allocations to various fishing gears and vessels, whereas spawning stock escapements of Pacific salmon (the number of fish "reserved" from the fisheries for spawning) are estimated primarily by summing up in-season fishing activity and stock run timings to build up a complete stock picture.

Developing a suite of alternative strategies is essential if the full range of dynamic changes in a fishery system is to be considered. Analytical models can take system uncertainty and errors in observation into account only if they can be programmed to consider all reasonable eventualities. If they are, the riskiness of each decision alternative can be assessed. Knowing the range of possible outcomes (risk assessment) for each alternative decision is essential for the management of risk (9).

The modeling exercise must also project anticipated performance measures arising from a decision, including performances of both stock and industry. It should generate a table of responses to be used if monitoring shows that actual performance measures will differ from anticipated ones. Monitoring and ongoing accountability of past decisions based on desirable performance over the planning period determine the effectiveness of the decision making strategy. Ongoing monitoring of decision performance with respect to objectives and continuous improvement over time are the ideas behind "total quality management" and "management by objectives" (10). When actual observations vary substantially from expected results, this "signal" should cause an adjustment in decision making and operational strategy. Systems that are "out of control" do not behave as expected over time, and these require quick adjustment in the model outlook and rapid evaluation of decision-making reliability.

Problem definition, development of alternatives, and evaluation require an integrated organization with the full contribution of all participants. Figure 13.1 illustrates an organizational process for fisheries management decisions with continual feedback. The circular process includes feedback from successive decisions over time taken from decision alternatives developed by all participants in the fishery and decisions made and translated into actual fisheries operations. Participants in decision analysis are drawn from all segments of the fishery system and together constitute a "management team." Team members represent stakeholders from four major groups (Figure 13.1):

- fisheries researchers, who are charged with leading biological research on the stock and who must define stock conservation constraints according to regular stock assessments and conduct research into linked economic and social aspects of the fisheries;
- the community, consisting of all individuals affected by the social and economic spin-offs of the fishing activity;
- fisheries operations, the group responsible for the industrial planning and market strategy, as well as with the implementation of decisions in the system that ensure that regulations implemented are enforced and monitored; and
- the fishing industry, consisting of all harvesters and processors and the organizations to which they belong insofar as these individuals and organizations participate in represented comanagement of the fishery and accept responsibility for maintaining a viable fishery sector within a sustainable fishery.

Among applied approaches, the "MACs" (management advisory committees) legislated by the Australian Fisheries Management Authority provide an effective model for responsible, team consensus management (11).

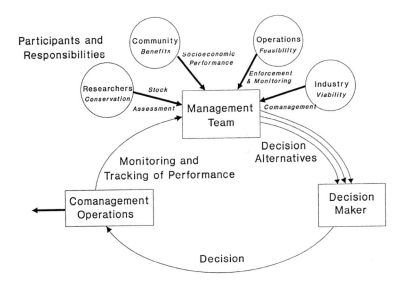

FIGURE 13.1.—Fisheries management decision process and supporting institutional arrangements.

Ideally, the management team, responsible for tasks of the decision analysis process, will include representatives from each of the participating groups—community leaders, biological and social scientists, and industry representatives (harvesting gears and processing sectors). The management team will be delegated the legal responsibility of acting in the best interests of the fishery to present decision alternatives to a separate decision-making body (as TACs are currently presented to a political leader), and to act directly in the ongoing seasonal operation and management of the fishery (e.g., to set local regulations, suballocate quotas, or prosecute offenders). It also will be the responsibility of the management team to track the performance of past decisions, by comparing the actual results of decision making with results anticipated when the decisions were taken, and incorporate this "wisdom" into future decisions.

It is only through a unified effort that all participants in a fishery system contribute to resolving decision problems related to management. Decision analysis provides a basis for structuring lines of communication between participants by requiring attention to specific problem elements in a structured framework. This framework then proceeds by developing and evaluating a suite of decision alternatives in relation to specified objectives, the ultimate resolution being effective decisions. Valuable lessons learned from decisions taken in the past are obtained by systematically comparing anticipated with actual performance of the system. This feedback is then used to adjust future decision making so that objectives will be reached over time (12).

NOTES

1. Various aspects of decision making in fisheries management have been explored by, among others, Rothschild (1973), MacKenzie (1974), Sinclair (1978), contributors to the book edited by Wooster (1988), and Hilborn and Walters (1992). Factors that complicate decision making were discussed by Ludwig et al. (1993), Smith et al. (1993), and Stephenson and Lane (1995).

2. Raiffa (1970) gave the fundamental concepts of decision analysis and Clemen (1991) presented its modern aspects. Bodily (1992) compiled current issues in the field and Lane (1989, 1992a, 1992b) addressed fisheries applications of decision analysis.
3. Clemen (1991, page 9).
4. For example, Halliday and Pinhorn (1995) and Rivard and Maguire (1993).
5. Jentoft (1989); Pinkerton and Weinstein (1995). See also Chapters 10–12 of this book.
6. Wooster (1988, page vii). Biological constraints have given rise to the "limit reference points" used in the precautionary approach to fisheries management.
7. Dunne (1990).
8. Larkin (1988, page 289).
9. The components of risk analysis—risk assessment and risk management—are discussed by Lane and Stephenson (1997).
10. Drucker (1954); Deming (1982).
11. Exel and Kaufmann (1997).
12. We are grateful for the helpful reviews of our manuscript provided by Jake Rice, Howard Powles, and an anonymous referee. The points of view expressed in this chapter are those of the authors and do not necessarily reflect those of any government agency. This work was funded by operating grants from the Natural Sciences and Engineering Research Council of Canada (OGP0043693, OGP0122822) to D. E. Lane.

BIBLIOGRAPHY

Acheson, J. M. 1975. The lobster fiefs: economic and ecological effects of territoriality in the Maine lobster fishery. Human Ecology 3:183–207.

Agardy, M. T. 1995. Critical area identification and zoning in coastal biosphere reserves: one way to make marine conservation work in Canada. Pages 214–219 *in* Shackell and Willison (1995).

Alexander, A. B., H. F. Moore, and W. C. Kendall. 1915. Otter-trawl fishery. Report of the U.S. Fisheries Commission for 1914, Appendix VI, Washington, D.C.

Allendorf, F. W., N. Ryman, and F. M. Utter. 1987. Genetics and fishery management: past, present, and future. Pages 1–19 *in* N. Ryman and F. Utter, editors. Population genetics and fishery management. University of Washington Press, Seattle.

Alpert, P. 1995. Incarnating ecosystem management. Conservation Biology 9:952–955.

Alverson, D. L., S. A. Murawski, and J. G. Pope. 1994. A global assessment of fisheries bycatch and discards. FAO (Food and Agriculture Organization of the United Nations) Fisheries Technical Paper 339.

Anderson, J. T., and E. L. Dalley. 1995. Spawning and recruitment of northern cod as measured by pelagic juvenile cod surveys following stock collapse, 1991–1994. Department of Fisheries and Oceans, Atlantic Fisheries Research Document 95/89, Dartmouth, Nova Scotia.

Anderson, R. W. 1987. "Bait up!" Dory fishing on Georges Bank. Pages 425–427 *in* R. Backus and D. Bourne, editors. Georges Bank. MIT Press, Cambridge, Massachusetts.

Angel, J. R., and five coauthors. 1994. Report of the workshop on Scotia-Fundy groundfish management from 1977 to 1993. Canadian Technical Report of Fisheries and Aquatic Sciences 1979.

Annand, C. 1992. Review of management measures for 1991 Scotia-Fundy groundfish fishery. CAFSAC (Canadian Atlantic Fisheries Scientific Advisory Committee) Research Document 92/103, Dartmouth, Nova Scotia.

Anonymous. 1983. Report of the working group on the methods of fish stock assessment. International Council for the Exploration of the Sea, C.M. 1983/Assess:17, Copenhagen.

Anonymous. 1987. Status of mixed species demersal finfish resources in New England and scientific basis for management. National Marine Fisheries Service, Northeast Fisheries Center, Reference Document N 87-08, Woods Hole, Massachusetts.

Anonymous. 1993. Status of fishery resources off the northeastern United States for 1993. NOAA (National Oceanic and Atmospheric Administration) Technical Memorandum NMFS-F/NEC-101, Washington, D.C.

Anonymous. 1995a. Compilation of the reports on the status of groundfish stocks of the Gulf of St. Lawrence. Department of Fisheries and Oceans, Atlantic Fisheries Stock Status Report 95/5, Dartmouth, Nova Scotia.

Anonymous. 1995b. Overview of the status of Canadian managed groundfish stocks in the Gulf of St. Lawrence and in the Canadian Atlantic. Department of Fisheries and Oceans, Atlantic Groundfish Overview, June 1995, Ottawa.

Anonymous. 1995c. Report on the status of Canadian managed groundfish stocks of the Newfoundland region. Department of Fisheries and Oceans, Atlantic Fisheries Stock Status Report 95/4, Dartmouth, Nova Scotia.

Anonymous. 1995d. Scotia-Fundy spring 1995 groundfish stock status report. Department of Fisheries and Oceans, Atlantic Fisheries Stock Status Report 95/6, Dartmouth, Nova Scotia.

Anthony, V. C. 1990. The New England groundfish fishery after 10 years under the Magnuson Fishery Conservation and Management Act. North American Journal of Fisheries Management 10:175–184.

Anthony, V. C. 1993. The state of groundfish resources off the northeastern United States. Fisheries 18(3):12–17.

Apostle, R., and six coauthors. In press. Community, market and state on the North Atlantic rim: challenges to modernity in the fisheries. University of Toronto Press, Toronto.

Armstrong, D. A., T. C. Wainwright, G. C. Jensen, P. A. Dinnel, and H. B. Andersen. 1993. Taking refuge from bycatch issues: red king crab (*Paralithodes camtschaticus*) and trawl fisheries in the eastern Bering Sea. Canadian Journal of Fisheries and Aquatic Sciences 50:1993–2000.

Atkinson, D. B. 1992. Some observations on the biomass and abundance of fish captured during stratified random bottom trawl surveys in NAFO divisions 2J3KL, fall 1981–91. Canadian Atlantic Fisheries Scientific Advisory Committee, Research Document 92/72, Dartmouth, Nova Scotia.

Auster, P. J., and eight coauthors. 1996. The impacts of mobile fishing gear on seafloor habitats in the Gulf of Maine (northwest Atlantic): implications for conservation of fish populations. Reviews in Fisheries Science 4:185–202.

Auster, P. J., and R. J. Malatesta. 1995. Assessing the role of non-extractive reserves for enhancing harvested populations in temperate and boreal marine systems. Pages 82–89 in Shackell and Willison (1995).

Auster, P. J., R. J. Malatesta, and C. L. S. Donaldson. 1997. Distributional responses to small-scale habitat variability by early juvenile silver hake, *Merluccius bilinearis*. Environmental Biology of Fishes 50:195–200.

Auster, P. J., R. J. Malatesta, and S. C. LaRosa. 1995. Patterns of microhabitat utilization by mobile megafauna on the southern New England (USA) continental shelf and slope. Marine Ecology Progress Series 127:77–85.

Auster, P. J., R. J Malatesta, S. C. LaRosa, R. A. Cooper, and L. L. Stewart. 1991. Microhabitat utilization by the megafaunal assemblage at a low relief outer continental shelf site—Middle Atlantic Bight, USA. Journal of Northwest Atlantic Fishery Science 11:59–69.

Bennett, B. A., and C. G. Attwood. 1991. Evidence for recovery of a surf-zone fish assemblage following the establishment of a marine reserve on the southern coast of South Africa. Marine Ecology Progress Series 75:173–181.

Berkes, F. 1987. Common property resource management and Cree Indian fisheries in subarctic Canada. Pages 66–91 in B. J. McCay and J. Acheson, editors. The question of the commons: the culture and ecology of communal resources. University of Arizona Press, Tucson.

Berkes, F. 1988. The intrinsic difficulty of predicting impacts: lessons from the James Bay hydro project. Environmental Impact Assessment Review 8:201–220.

Berkes, F. 1993. Traditional ecological knowledge in perspective. Pages 1–10 in J. T. Inglis, editor. Traditional ecological knowledge: concepts and cases. International Development Research Centre, Ottawa.

Beverton, R. J. H., and S. J. Holt. 1957. On the dynamics of exploited fish populations. Fisheries Investigations Series II, Marine Fisheries, Great Britain Ministry of Agriculture, Fisheries and Food 19.

Bishop, C. A., and eight coauthors. 1994. An assessment of the cod stock in NAFO divisions 2J+3KL. Northwest Atlantic Fisheries Organization, Scientific Council Research Document 94/40, Dartmouth, Nova Scotia.

Bodily, S. E., editor. 1992. Decision and risk analysis. Interfaces 22(6).

Bohnsack, J. A. 1992. Reef resource habitat protection: the forgotten factor. Pages 117–129 *in* R. H. Stroud, editor. Stemming the tide of coastal fish habitat loss. National Coalition for Marine Conservation, Savannah, Georgia.

Bonin, J. P., D. C. Jones, and L. Putterman. 1993. Theoretical and empirical studies of producer cooperatives: will ever the twain meet? Journal of Economic Literature 31:1290–1320.

Bourne, D. W. 1987. The fisheries. *In* R. H. Backus, editor. Georges Bank. MIT Press, Cambridge, Massachusetts.

Bowditch, N. 1995. The American practical navigator. U.S. Defense Mapping Agency, Hydrographic/Topographic Center, Publication 9, Bethesda, Maryland.

Brander, K. M. 1995. The effects of temperature on growth of Atlantic cod (*Gadus morhua* L.). ICES Journal of Marine Science 52:1–10.

Brodie, W. B. 1987. American plaice in divisions 3LNO—an assessment update. Northwest Atlantic Fisheries Organization, Scientific Council Research Document 87/40, Dartmouth, Nova Scotia.

Brown, B. E., J. A. Brennan, E. G. Heyerdahl, M. D. Grosslein, and R. C. Hennemuth. 1976. The effect of fishing on the marine finfish biomass of the northwest Atlantic from the Gulf of Maine to Cape Hatteras. International Commission for the Northwest Atlantic Fisheries, Research Bulletin 12:49–68.

Brown, B. E., and R. G. Halliday. 1983. Fisheries resources of the northwest Atlantic—some responses to extreme fishing perturbations. Pages 96–109 *in* Proceedings of the Joint Oceanographic Assembly 1982—general symposia. Canadian National Committee, Scientific Committee on Oceanic Research, Ottawa.

Brownstein, J., and J. Tremblay. 1994. Traditional property rights and cooperative management in the Canadian lobster fishery. The Lobster Newsletter 7:5.

Buch, E., S. A. Horsted, and H. Hovgård. 1994. Fluctuations in the occurrence of cod in Greenland waters and their possible causes. ICES Marine Science Symposium 198:158–174.

Bugden, G. L. 1991. Changes in temperature–salinity characteristics of the deeper waters of the Gulf of St. Lawrence over the past several decades. Pages 139–147 *in* J.-C. Therriault, editor. The Gulf of St. Lawrence: small ocean or big estuary? Canadian Special Publication of Fisheries and Aquatic Sciences 113.

Burke, L., and seven coauthors. 1994. The Scotia-Fundy inshore dragger fleet ITQ program. Background, implementation, and results to date. International Council for the Exploration of the Sea, C.M. 1994/T:35, Copenhagen.

Campana, S. E., R. K. Mohn, S. J. Smith, and G. Chouinard. 1995. Spatial visualization of a temperature-based growth model for Atlantic cod (*Gadus morhua*) off the eastern coast of Canada. Canadian Journal of Fisheries and Aquatic Sciences 52:2445–2456.

Campbell, A., and D. S. Pezzack. 1986. Relative egg production and abundance of berried lobsters, *Homarus americanus*, in the Bay of Fundy and off southwestern Nova Scotia. Canadian Journal of Fisheries and Aquatic Sciences 43:2190–2196.

Carr, H. A., and P. Caruso. 1992. Application of a horizontal separating panel to reduce bycatch in the small mesh whiting fishery. Proceedings of the Marine Technical Society 1:401–407.

Carr, M. H., and D. C. Reed. 1993. Conceptual issues relevant to marine harvest refuges: examples from temperate reef fishes. Canadian Journal of Fisheries and Aquatic Sciences 50:2019–2028.

CDFG (California Department of Fish and Game). 1993. Final environmental impact statement: Marine Resources Protection Act of 1990, ecological reserves. CDFG, Sacramento.

Chapman, D. C., and R. C. Beardsley. 1989. On the origin of shelf water in the Middle Atlantic Bight. Journal of Physical Oceanography 19:384–391.

Charles, A. T. 1994. Towards sustainability: the fishery experience. Ecological Economics 11:201–211.

Chouinard, G. A., and A. Fréchet. 1994. Fluctuations in the cod stocks of the Gulf of St. Lawrence. ICES Marine Science Symposium 198:121–139.

Christensen, N. L., and 12 coauthors. 1996. The report of the Ecological Society of America Committee on the Scientific Basis for Ecosystem Management. Ecological Applications 6:665–691.

Christy, F. T. 1982. Territorial use rights in marine fisheries: definitions and conditions. FAO (Food and Agriculture Organization of the United Nations) Fisheries Technical Paper 227.

Clark, S. H., and B. E. Brown. 1977. Changes in biomass of finfish and squids from the Gulf of Maine to Cape Hatteras, 1963–1974, as determined from research vessel survey data. U.S. National Marine Fisheries Service Fishery Bulletin 75:1–21.

Clark, S. H., W. J. Overholtz, and R. C. Hennemuth. 1982. Review and assessment of the Georges Bank and Gulf of Maine haddock fishery. Journal of Northwest Atlantic Fishery Science 3:1–27.

Clemen, R. T. 1991. Making hard decisions: an introduction to decision analysis. PWS-Kent, Boston.

Cohen, E. B., D. G. Mountain, and R. N. O'Boyle. 1991. Local-scale versus large-scale factors affecting recruitment. Canadian Journal of Fisheries and Aquatic Sciences 48:1003–1006.

Cohen, F. G. 1986. Treaties on trial; the continuing controversy over northwest Indian fishing rights. University of Washington Press, Seattle.

Colbourne, E. 1995. Oceanographic conditions and climate change in the Newfoundland region during 1994. Department of Fisheries and Oceans, Atlantic Fisheries Research Document 95/3, Dartmouth, Nova Scotia.

Colbourne, E., S. Narayanan, and S. Prinsenberg. 1994. Climatic changes and environmental conditions in the northwest Atlantic, 1970–1993. ICES Marine Science Symposium 198:311–322.

Colton, J. B., Jr. 1959. A field observation of mortality of marine fish larvae due to warming. Limnology and Oceanography 4:219–222.

Colton, J. B., Jr. 1968. Recent trends in subsurface temperature in the Gulf of Maine and contiguous waters. Journal of the Fisheries Research Board of Canada 26:2427–2437.

Colton, J. B., Jr. 1972. Temperature trends and the distribution of groundfish in continental shelf waters, Nova Scotia to Long Island. U.S. National Marine Fisheries Service Fishery Bulletin 70:637–657.

Copes, P. 1986. A critical review of the individual quota as a device in fisheries management. Land Economics 62:278–291.

Creed, C. F. 1996. Social responses to ITQs: cheating and stewardship in the Canadian Scotia-Fundy inshore mobile gear sector. Pages 72–83 *in* D. L. Burke, R. N. O'Boyle, P. Partington, and M. Sinclair, editors. Report of the second workshop on Scotia-Fundy groundfish management. Canadian Technical Report of Fisheries and Aquatic Sciences 2100.

Crowley, R. W., and H. Palsson. 1992. Rights based fisheries management in Canada. Marine Resource Economics 7:1–21.

Daan, N., and M. P. Sissenwine, editors. 1991. Multispecies models relevant to the management of living resources. ICES Marine Science Symposium 193.

Davis, A. 1984. Property rights and access management in the small boat fishery: a case study from southwest Nova Scotia. Pages 133–164 in C. Lamson and A. J. Hanson, editors. Atlantic fisheries and coastal communities: fisheries decision-making case studies. Dalhousie Ocean Studies Program, Halifax, Nova Scotia.

Day, D. 1995. Tending the Achilles' heel of NAFO: Canada acts to protect the Nose and Tail of the Grand Banks. Marine Policy 19:257–270.

Dayton, P. K., S. F. Thrush, M. T. Agardi, and R. J. Hofman. 1995. Environmental effects of marine fishing. Aquatic Conservation: Marine and Freshwater Ecosystems 5:205–232.

DeAlteris, J. T., K. Castro, and S. A. Testaverde. 1990. Effects of mesh size in the body of a bottom trawl on the catch retained in the cod end. Pages 60–70 in J. T. DeAlteris and M. Grady, editors. Proceedings of the Fishery Conservation Engineering Workshop. Rhode Island Sea Grant, Kingston.

DeAlteris, J. T., H. Milliken, and D. Morse. In press. Bycatch reduction in the northwest Atlantic small mesh bottom trawl fishery for silver hake (*Merluccius bilinearis*). In Proceedings of the Second World Fisheries Congress, Brisbane, Australia.

DeAlteris, J. T., and D. M. Reifsteck. 1993. Escapement and survival of fish from the codend of a demersal trawl. ICES Marine Science Symposium 196:128–131.

de Cárdenas, E. 1996. Some considerations about annual growth rate variations in cod stocks. Northwest Atlantic Fisheries Organization, Scientific Council Studies 24:97–107.

deYoung, B., and G. A. Rose. 1993. On recruitment and distribution of Atlantic cod (*Gadus morhua*) off Newfoundland. Canadian Journal of Fisheries and Aquatic Sciences 50:2729–2741.

Deming, W. E. 1982. Quality, productivity, and competitive position. MIT Center for Advanced Engineering Study, Cambridge, Massachusetts.

Deriso, R. B. 1982. Relationship of fishing mortality to natural mortality and growth at the level of maximum sustainable yield. Canadian Journal of Fisheries and Aquatic Sciences 39:1054–1058.

Deriso, R. B. 1987. Optimum $F_{0.1}$ criteria and their relationship to maximum sustainable yield. Canadian Journal of Fisheries and Aquatic Sciences 44(Supplement 2):339–348.

DFO (Department of Fisheries and Oceans). 1989. The Scotia-Fundy lobster fishery—phase one: issues and considerations—summary report. DFO, Halifax, Nova Scotia.

Dickson, R. R., J. Meincke, S.-A. Malmberg, and A. J. Lee. 1988. The great salinity anomaly in the northern North Atlantic, 1968–82. Progress in Oceanography 20:103–151.

Doeringer, P., P. Moss, and D. Terkla. 1986. The New England fishing economy. University of Massachusetts Press, Amherst.

Doulman, D. J. 1993. Community-based fishery management: towards the restoration of traditional practices in the South Pacific. Marine Policy 17:109–117.

Drinkwater, K. F. 1987. "Sutcliffe revisited": previously published correlations between fish stocks and environmental indices and their recent performance. Canadian Technical Report of Fisheries and Aquatic Sciences 1556:41–61.

Drinkwater, K. F. 1996. Atmospheric and oceanic variability in the northwest Atlantic during the 1980s and early-1990s. Journal of Northwest Atlantic Fishery Science 18:77–97.

Drinkwater, K. F., E. Colbourne, and D. Gilbert. 1996. Overview of environmental conditions in the northwest Atlantic in 1994. Northwest Atlantic Fisheries Organization, Scientific Council Studies 25:25–58.

Drinkwater, K. F., and R. A. Myers. 1987. Testing predictions of marine fish and shellfish landings from environmental variables. Canadian Journal of Fisheries and Aquatic Sciences 44:1568–1573.

Drucker, P. F. 1954. The practice of management. Harper & Row, New York.

Dugan, J. E., and G. E. Davis. 1993. Applications of marine refugia to coastal fisheries management. Canadian Journal of Fisheries and Aquatic Sciences 50:2029–2042.

Dunne, E. 1990. Report of the implementation task force on northern cod. Department of Fisheries and Oceans, Atlantic Fisheries Adjustment Program, Ottawa.

Dutil, J.-D., Y. Lambert, G. A. Chouinard, and A. Fréchet. 1995. Fish condition: what should we measure in cod (*Gadus morhua*). Department of Fisheries and Oceans, Atlantic Fisheries Research Document 95/11, Dartmouth, Nova Scotia.

Dyer, C. L., and J. R. McGoodwin, editors. 1994. Folk management in the world's fisheries: lessons for modern fisheries management. University Press of Colorado, Boulder.

Exel, M., and B. Kaufmann. 1997. Allocation of fishing rights: implementation issues in Australia. Pages 246–255 *in* E. K. Pikitch, D. D. Huppert, and M. P. Sissenwine, editors. Global trends: fisheries management. American Fisheries Society Symposium 20, Bethesda, Maryland.

Fanning, P., K. Zwanenburg, and M. A. Showell. 1987. Haddock nursery closed areas: delineation and impact. CAFSAC (Canadian Atlantic Fisheries Scientific Advisory Committee) Research Document 87/59, Dartmouth, Nova Scotia.

FAO (Food and Agriculture Organization of the United Nations). 1994. Review of the state of world marine fishery resources. FAO Fisheries Technical Paper 335.

Felt, L. 1990. Obstacles to user-participation in the Canadian Atlantic salmon fishery. Marine Policy 14:345–360.

Felt, L. 1994. Two tales of a fish: the social construction of indigenous knowledge among Atlantic Canadian salmon fishers. Pages 251–286 *in* C. L. Dyer and J. R. McGoodwin, editors. Folk management in the world's fisheries: lessons for modern fisheries management. University Press of Colorado, Boulder.

Finlayson, A. C. 1994. Fishing for truth: a sociological analysis of northern cod stock assessments from 1977 to 1990. Institute of Social and Economic Research, Memorial University, St. John's, Newfoundland.

FMS (Fisheries and Marine Service). 1976. Policy for Canada's commercial fisheries. Environment Canada Miscellaneous Publication, Ottawa.

Fox, W. W. 1975. Fitting the generalized stock production model by least squares and equilibrium approximation. U.S. National Marine Fisheries Service Fishery Bulletin 73:23–37.

Frank, K. T., K. F. Drinkwater, and F. H. Page. 1994. Possible causes of recent trends and fluctuations in Scotian Shelf/Gulf of Maine cod stocks. ICES Marine Science Symposium 198:110–120.

Frank, K. T., J. Simon, and J. E. Carscadden. 1996. Recent excursions of capelin (*Mallotus villosus*) to the Scotian Shelf and Flemish Cap during anomalous hydrographic conditions. Canadian Journal of Fisheries and Aquatic Sciences 53:1473–1486.

Fréchet, A. 1991. A declining cod stock in the Gulf of St. Lawrence: how can we learn from the past? NAFO (Northwest Atlantic Fisheries Organization) Scientific Council Studies 16:95–102.

Freeman, M. 1989a. Graphs and gaffs: a cautionary tale in the common-property resources debate. Pages 99–109 *in* F. Berkes, editor. Common property resources: ecology and community-based sustainable development. Belhaven Press, London.

Freeman, M. 1989b. The Alaska Eskimo Whaling Commission: successful co-management under extreme conditions. Pages 137–153 *in* E. Pinkerton, editor. Cooperative management of local fisheries: new directions for improved management and community development. University of British Columbia Press, Vancouver.

Gabriel, W. L. 1985. Spawning stock biomass per recruit analysis for seven northwest Atlantic demersal finfish species. National Marine Fisheries Service, Northeast Fisheries Center, Reference Document N 85-04, Woods Hole, Massachusetts.

Gabriel, W. L. 1992. Persistence of demersal fish assemblages between Cape Hatteras and Nova Scotia, northwest Atlantic. Journal of Northwest Atlantic Fishery Science 14:29–46.

Gabriel, W. L., M. P. Sissenwine, and W. J. Overholtz. 1989. Analysis of spawning stock biomass per recruit: an example for Georges Bank haddock. North American Journal of Fisheries Management 9:383–391.

Gadgil, M., and F. Berkes. 1991. Traditional resource management systems. Resource Management and Optimization 18:127–141.

Gadgil, M., F. Berkes, and C. Folke. 1993. Indigenous knowledge for conservation of biodiversity. Ambio 22:151–156.

Gardner, M. 1988. Enterprise allocation system in the offshore groundfish sector in Atlantic Canada. Marine Resource Economics 5:389–414.

Garrett, C. J. R., J. R. Keeley, and D. A. Greenberg. 1978. Tidal mixing versus thermal stratification in the Bay of Fundy and Gulf of Maine. Atmosphere-Ocean 16:403–423.

German, A. W. 1987. History of the early fisheries: 1720–1930. Pages 409–424 *in* R. H. Backus, editor. Georges Bank. MIT Press, Cambridge, Massachusetts.

Gerrior, P., F. M. Serchuk, and K. C. Mays. 1994. How mixed is the mixed species trawl fishery on Georges Bank or evaluating fishery performance via an observer program. International Council for the Exploration of the Sea, C.M. 1994/G:11, Copenhagen.

Gilbert, D., and B. Pettigrew. 1997. A study of the interannual variability of the CIL core temperature in the Gulf of St. Lawrence. Canadian Journal of Fisheries and Aquatic Sciences 54 (Supplement 1):57–67.

Gomes, M. C., and R. L. Haedrich. 1992. Predicting community dynamics from food web structure. Pages 277–293 *in* G. T. Rowe and V. Pariente, editors. Deep-sea food chains and the global carbon cycle. Kluwer, Dordrecht, The Netherlands.

Gomes, M. C., R. L. Haedrich, and J. C. Rice. 1992. Biogeography of groundfish assemblages on the Grand Bank. Journal of Northwest Atlantic Fishery Science 14:13–27.

Gomes, M. C., R. L. Haedrich, and M. G. Villagarcia. 1995. Spatial and temporal changes in the groundfish assemblages on the northeast Newfoundland/Labrador shelf, 1978–1991. Fisheries Oceanography 4:85–101.

Goode, G. B., editor. 1887. The fisheries and fisheries industries of the United States, section V: history and methods of the fisheries, volumes 1 and 2. Government Printing Office, Washington, D.C.

Goodyear, C. P. 1993. Spawning stock biomass per recruit in fisheries management: foundation and current use. Pages 67–81 *in* Smith et al. (1993).

Gordon, H. S. 1954. The economic theory of a common property resource: the fishery. Journal of Political Economy 62:124–142.

Gotceitas, V., and J. A. Brown. 1993. Substrate selection by juvenile Atlantic cod (*Gadus morhua*): effects of predation risk. Oecologia 93:31–37.

Gotceitas, V., J. A. Brown, and S. Mercer. 1994. Laboratory investigations on substrate use by juvenile Atlantic cod. Pages 92–96 *in* D. Stevenson and E. Braasch, editors. Gulf of Maine habitat: workshop proceedings. University of New Hampshire, Sea Grant Program, Report UNHMP-T/DR-SG-94-18, Durham.

Goudey, C. A. 1995. The experimental pair trawl fishery for tuna in the northwest Atlantic. MIT Sea Grant College Program, Technical Report MITSG 96-17, Cambridge, Massachusetts.

Gough, J. 1991. Fisheries management in Canada 1880–1910. Canadian Manuscript Report of Fisheries and Aquatic Sciences 2105.

Graham, H. W. 1952. Mesh regulation to increase the yield of the Georges Bank haddock fishery. Pages 23–33 *in* International Commission for the Northwest Atlantic Fisheries, Annual Report 2, Dartmouth, Nova Scotia.

Graham, H. W., and E. D. Premetz. 1955. First year of mesh regulation in the Georges Bank haddock fishery. U.S. Fish and Wildlife Service, Special Scientific Report Fisheries 142.

Greenberg, D. A. 1983. Modeling the mean barotropic circulation in the Bay of Fundy and Gulf of Maine. Journal of Physical Oceanography 13:886–904.

Grumbine, R. E. **1994.** What is ecosystem management? Conservation Biology 8:27–38.
Gulland, J. A., and L. K. Boerema. **1973.** Scientific advice on catch levels. U.S. National Marine Fisheries Service Fishery Bulletin 71:325–335.
Haché, J. E., chairman. **1989.** Report of the Scotia-Fundy Groundfish Task Force. Department of Fisheries and Oceans, Fs 23-157/1989E, Ottawa.
Hachey, H. B., F. Hermann, and W. B. Bailey. **1954.** The waters of the ICNAF convention area. International Commission for the Northwest Atlantic Fisheries, Annual Proceedings 4:67–102, Dartmouth, Nova Scotia.
Haedrich, R. L., and J. Fischer. **1996.** Stability and change of exploited fish communities in a cold ocean continental shelf ecosystem. Senckenbergiana Maritima 27:237–243.
Haedrich, R. L., M. G. Villagarcia, and M. C. Gomes. **1995.** Scale of marine protected areas on Newfoundland's continental shelf. Pages 48–53 *in* Shackell and Willison (1995).
Halliday, R. G. **1988.** Use of seasonal spawning area closures in the management of haddock fisheries in the northwest Atlantic. Northwest Atlantic Fisheries Organization Scientific Council Studies 12:27–36.
Halliday, R. G. **1987.** Haddock spawning area closures in the northwest Atlantic, 1970–1987. Northwest Atlantic Fisheries Organization Scientific Council Research Document 87/13, Dartmouth, Nova Scotia.
Halliday, R. G., F. G. Peacock, and D. L. Burke. **1992.** Development of management measures for the groundfish fishery in Atlantic Canada: a case study of the Nova Scotia inshore fleet. Marine Policy 16:411–426.
Halliday, R. G., and A. T. Pinhorn. **1985.** Present management strategies in Canadian Atlantic marine fisheries, their rationale and the historical context in which their usage developed. Pages 10–33 *in* R. Mahon, editor. Towards the inclusion of fishery interactions in management advice. Canadian Technical Report of Fisheries and Aquatic Sciences 1347.
Hammill, M. O., and B. Mohn. **1994.** A model of grey seal predation on Atlantic cod on the Scotian Shelf and Gulf of St. Lawrence. Department of Fisheries and Oceans, Atlantic Fisheries Research Document 94/75, Dartmouth, Nova Scotia.
Hanson, J. M., and G. A. Chouinard. **1992.** Evidence that size-selective mortality affects growth of Atlantic cod (*Gadus morhua* L.) in the southern Gulf of St. Lawrence. Journal of Fish Biology 41:31–41.
Harris, L. **1990.** Independent review of the state of the northern cod stock. Department of Fisheries and Oceans, Fs 23-160/1990E, Ottawa.
Harris, H. J., S. Richman, V. A. Harris, and C. J. Yarbrough. **1987.** Coupling ecosystem science with management: a Great Lakes perspective from Green Bay, Lake Michigan, USA. Environmental Management 11:619–625.
Hennemuth, R. C., and S. Rockwell. **1987.** History of fisheries conservation and management. Pages 430–446 *in* R. H. Backus, editor. Georges Bank. MIT Press, Cambridge, Massachusetts.
Herrington, W. C. **1935.** Modifications in gear to curtail the destruction of undersized fish in otter trawling. U.S. Bureau of Fisheries, Investigation Report 24, Washington, D.C.
Hilborn, R., and C. J. Walters. **1992.** Quantitative fisheries stock assessment: choice, dynamics, and uncertainty. Chapman and Hall, New York.
Holden, M. J. **1971.** Report of the ICES/ICNAF working groups on selectivity analysis. International Council for the Exploration of the Sea Cooperative Research Report, Series A, 25.
Holling, C. S. **1978.** Adaptive environmental assessment and management. Wiley, New York.
Howell, W. H., and R. Langan. **1987.** Commercial trawler discards of four flounder species in the Gulf of Maine. North American Journal of Fisheries Management 7:6–17.
Hutchings, J. A. **1995.** Seasonal marine protected areas within the context of spatio-temporal variation in the northern cod fishery. Pages 39–47 *in* Shackell and Willison (1995).

Hutchings, J. A., and R. A. Myers. 1994a. Timing of cod reproduction: interannual variability and the influence of temperature. Marine Ecology Progress Series 108:21–31.

Hutchings, J. A., and R. A. Myers. 1994b. What can be learned from the collapse of a renewable resource? Atlantic cod, *Gadus morhua*, off Newfoundland and Labrador. Canadian Journal of Fisheries and Aquatic Sciences 51:2126–2146.

Hutchings, J. A., R. A. Myers, and G. R. Lilly. 1993. Geographic variation in the spawning of Atlantic cod, *Gadus morhua*, in the northwest Atlantic. Canadian Journal of Fisheries and Aquatic Sciences 50:2457–2467.

ICES (International Council for the Exploration of the Sea). 1994. Joint report of the ICES advisory committee on fishery management and the advisory committee on the marine environment, 1994. International Council for the Exploration of the Sea Cooperative Research Report 203, Copenhagen.

ICNAF (International Commission for the Northwest Atlantic Fisheries). 1974. Report of assessments subcommittee. ICNAF Annual Proceedings, Appendix I, Dartmouth, Nova Scotia.

ICNAF (International Commission for the Northwest Atlantic Fisheries). 1962–1972. Fishery statistics for the years 1960–1970. ICNAF Statistical Bulletin, volumes 10–20, Dartmouth, Nova Scotia.

IPHC (International Pacific Halibut Commission). 1987. The Pacific halibut: biology, fishery and management. IPHC Technical Report 22.

Jakobsen, T. 1993. The behavior of F_{low}, F_{med}, and F_{high} in response to variation in parameters used for their estimation. Pages 119–125 *in* Smith et al. (1993).

Jefferson, T. A., S. Leatherwood, and M. A. Webber. 1993. FAO species identification guide. FAO (Food and Agriculture Organization of the United Nations), Rome.

Jentoft, S. 1989. Fisheries co-management: delegating government responsibility to fishermen's organizations. Marine Policy 13:137–154.

Jentoft, S., and T. Kristoffersen. 1989. Fishermen's co-management: the case of the Lofoton fishery. Human Organizations 48:355–365.

Jentoft, S., and B. J. McCay. 1995. User participation in fisheries management: lessons drawn from international experiences. Marine Policy 19:227–246.

Johannes, R. E. 1993. Integrating traditional ecological knowledge and management with environmental assessment. Pages 33–40 *in* J. T. Inglis, editor. Traditional knowledge: concepts and cases. International Development Research Centre, Ottawa.

Johannes, R. E. 1981. Words of the lagoon: fishing and marine lore in the Palau District of Micronesia. University of California Press, Berkeley.

Johannes, R. E. 1978. Traditional marine conservation methods in Oceania, and their demise. Annual Review of Ecology and Systematics 9:349–364.

Johnson, M., editor. 1992. Lore: capturing traditional environmental knowledge. International Development Research Centre, Dene Cultural Institute, Hay River, Northwest Territories.

Jones, J. B. 1992. Environmental impact of trawling on the seabed: a review. Australian Journal of Marine and Freshwater Research 26:59–67.

Kearney, J. 1985. The transformation of the Bay of Fundy herring fisheries in 1976–1987: an experiment in fisherman-government co-management. Pages 165–203 *in* C. Lamson and A. Hanson, editors. Atlantic fisheries and coastal communities: fisheries decision-making case studies. Institute of Resource and Environmental Studies, Dalhousie University, Halifax, Nova Scotia.

Kenney, J. F., A. J. Blott, and V. E. Nulk. 1992. Experiments with a Nordmore grate in the Gulf of Maine shrimp fishery. Report of the New England Fishery Management Council to the National Oceanic and Atmospheric Administration, Award NA87EA-H-00052, Washington, D.C.

Kent, G. 1987. Fish, food and hunger: the potential of fisheries for alleviating malnutrition. Westview Press, Boulder, Colorado.

Kirby, M. J. L. 1982. Navigating troubled waters, a new policy for the Atlantic fisheries. Report of the Task Force on Atlantic Fisheries. Supply and Services, CP 32-43/1983-1E, Ottawa.

Klima, E. F., G. A. Matthews, and F. J. Patella. 1986. Synopsis of the Tortugas pink shrimp fishery, 1960–1983, and the impact of the Tortugas sanctuary. North American Journal of Fisheries Management 6:301–310.

Kloppenburg, J. 1991. Social theory and the de/reconstruction of agricultural science: local knowledge for an alternative agriculture. Rural Sociology 56:519–548.

Koslow, J. A. 1984. Recruitment patterns in northwest Atlantic fish stocks. Canadian Journal of Fisheries and Aquatic Sciences 41:1722–1729.

Koslow, J. A., K. R. Thompson, and W. Silvert. 1987. Recruitment to northwest Atlantic cod (*Gadus morhua*) and haddock (*Melanogrammus aeglefinus*) stocks: influence of stock size and climate. Canadian Journal of Fisheries and Aquatic Sciences 44:26–39.

Koutitonsky, V. G., and G. L. Bugden. 1991. The physical oceanography of the Gulf of St. Lawrence: a review with emphasis on the synoptic variability of the motion. Pages 57–90 *in* J.-C. Therriault, editor. The Gulf of St. Lawrence: small ocean or big estuary? Canadian Special Publication of Fisheries and Aquatic Sciences 113.

Lackey, R. T. In press. Seven pillars of ecosystem management. Landscape and Urban Planning.

Lane, D. E. 1989. Operational research and fisheries management. European Journal of Operational Research 42:229–242.

Lane, D. E. 1992a. Management science in the control and management of fisheries: an annotated bibliography. American Journal of Mathematics and Management Science 12:101–152.

Lane, D. E. 1992b. Regulation of commercial fisheries in Atlantic Canada: a decision perspective. Optimum 23(2):37–50.

Lane, D. E., and R. L. Stephenson. 1995. A decision making framework for the development of management plans. Department of Fisheries and Oceans, Atlantic Fisheries Research Document 95/80, Dartmouth, Nova Scotia.

Lane, D. E., and R. L. Stephenson. 1997. A framework for risk analysis in fisheries decision-making. ICES Journal of Marine Science 54.

Langton, R. W. 1982. Diet overlap between Atlantic cod, *Gadus morhua*, silver hake, *Merluccius bilinearis*, and fifteen other northwest Atlantic finfish. U.S. National Marine Fisheries Service Fishery Bulletin 80:745–759.

Langton, R. W., P. J. Auster, and D. C. Schneider. 1995. A spatial and temporal perspective on research and management of groundfish in the northwest Atlantic. Reviews in Fisheries Science 3:201–229.

Langton, R. W., R. S. Steneck, V. Gotceitas, F. Juanes, and P. Lawton. 1996. The interface between fisheries research and habitat management. North American Journal of Fisheries Management 16:1–7.

Langton, R. W., and L. Watling. 1990. The fish-benthos connection: a definition of prey groups in the Gulf of Maine. Pages 424–438 *in* M. Barnes and R. N. Gibson, editors. Trophic relationships in the marine environment. Aberdeen University Press, Aberdeen, UK.

Larkin, P. 1988. Comments on the workshop presentations. Pages 287–289 *in* W. S. Wooster, editor. Fishery science and management: objectives and limitations. Lecture notes on coastal and estuarine studies, volume 28. Springer-Verlag, New York.

Larkin, P. A. 1977. An epitaph for the concept of maximum sustained yield. Transactions of the American Fisheries Society 106:1–11.

Lauzier, L. M. 1965. Long-term temperature variations in the Scotian Shelf area. International Commission for the Northwest Atlantic Fisheries, Special Publication 6:807–816, Dartmouth, Nova Scotia.

Lauzier, L. M., and R. W. Trites. 1958. The deep waters in the Laurentian Channel. Journal of the Fisheries Research Board of Canada 15:1247–1257.

Lazar, N. 1993. Analysis of the sink gillnet fishery in the Gulf of Maine and adjacent waters. Master's thesis. University of Rhode Island, Kingston.

Lazier, J. R. N., and D. G. Wright. 1993. Annual velocity variations in the Labrador Current. Journal of Physical Oceanography 23:659–678.

Lee, K. N. 1989. The Columbia River basin: experimenting with sustainability. Environment 31(6):6–11, 30–33.

Lien, J., W. Barney, S. Todd, R. Seton, and J. Guzzwell. 1992. Effects of adding sounds to cod traps on the probability of collisions by humpback whales. Pages 701–708 in J. Thomas, editor. Marine mammal sensory systems. Plenum, New York.

Lilly, G. R. 1994. Predation by Atlantic cod on capelin on the southern Labrador and northeast Newfoundland shelves during a period of changing spatial distributions. ICES Marine Science Symposium 198:600–611.

Lilly, G. R. 1995. Did the feeding of the cod off southern Labrador and eastern Newfoundland decline in the 1990s? Department of Fisheries and Oceans, Atlantic Fisheries Research Document 95/74, Dartmouth, Nova Scotia.

Lilly, G. R., H. Hop, D. E. Stansbury, and C. A. Bishop. 1994. Distribution and abundance of polar cod (*Boreogadus saida*) off southern Labrador and eastern Newfoundland. International Council for the Exploration of the Sea, C.M. 1994/O:6, Copenhagen.

Loder, J. W. 1980. Topographic rectifications of tidal currents on the sides of Georges Bank. Journal of Physical Oceanography 10:1399–1416.

Loder, J. W., B. Petrie, and G. Gawarkiewicz. In press. The coastal ocean off northeastern North America: a large-scale view. In A. R. Robinson and K. H. Brink, editors. The sea, volume 11. The global coastal ocean: regional studies and synthesis. Wiley, New York.

Lough, R. G., and nine coauthors. 1994. Influence of wind-driven advection on interannual variability in cod egg and larval distributions on Georges Bank: 1982 vs 1985. ICES Marine Science Symposium 198:356–378.

Lough, R. G., and six coauthors. 1989. Ecology and distribution of juvenile cod and haddock in relation to sediment type and bottom currents on Georges Bank. Marine Ecology Progress Series 56:1–12.

Ludwig, D., R. Hilborn, and C. Walters. 1993. Uncertainty, resource exploitation, and conservation: lessons from history. Science 260:17, 36.

Mace, P. M. 1993. Will private owners practice prudent resource management? Fisheries 18(9):29–31.

Mace, P. M., and M. P. Sissenwine. 1993. How much spawning per recruit is enough? Pages 101–118 in Smith et al. (1993).

MacInnes, D., and A. Davis. 1996. Representational management or management of representation?: the place of fishers in Atlantic Canadian fisheries management. Pages 317–332 in R. M. Meyer and five coeditors. Fisheries resource utilization and policy. Proceedings of the World Fisheries Congress, Theme 2. Oxford & IBH, New Delhi.

MacKenzie, W. C. 1974. Conceptual aspects of strategic planning for fishery management and development. Journal of the Fisheries Research Board of Canada 31:1705–1712.

MacLennan, D. N., and E. J. Simmonds. 1994. Fisheries acoustics. Chapman and Hall, London.

Maguire, J.-J., and P. M. Mace. 1993. Biological reference points for Canadian Atlantic gadoid stocks. Pages 321–331 in Smith et al. (1993).

Maguire, J.-J., B. Neis, and P. Sinclair. 1995. What are we managing anyway?: the need for an interdisciplinary approach to managing fisheries ecosystems. Dalhousie Law Review 18:141–153.

Mahon, R., and R. W. Smith. 1989. Demersal fish assemblages on the Scotian Shelf, northwest Atlantic: spatial distribution and persistence. Canadian Journal of Fisheries and Aquatic Sciences 46(Supplement 1):134–152.

Mailhot, J. 1993. Traditional ecological knowledge: the diversity of knowledge systems and their study. The Great Whale Public Review Support Office, Montreal.

Malatesta, R. J., P. J. Auster, and S. C. LaRosa. 1994. Temporal variation in microhabitat use by mobile fauna (Middle Atlantic Bight, USA): an experimental approach. Pages 41–58 in Proceedings of the American Academy of Underwater Sciences, Costa Mesa, California.

Man, A., R. Law, and N. V. C. Polunin. 1995. Role of marine reserves in recruitment to reef fisheries: a metapopulation model. Biological Conservation 71:197–204.

Mann, K. H. 1993. Physical oceanography, food chains, and fish stocks: a review. ICES Journal of Marine Science 50:105–119.

Manning, J. 1991. Middle Atlantic Bight salinity: interannual variability. Continental Shelf Research 11:123–137.

Marchesseault, G., R. Ruais, and S. Wang. 1980. History and status of the Atlantic demersal finfish fishery management plan. NOAA (National Oceanic and Atmospheric Administration) Technical Memorandum NMFS-F/NEC-2, Woods Hole, Massachusetts.

Marko, J. R., D. B. Fissel, P. Wadhams, P. M. Kelly, and R. D. Brown. 1994. Iceberg severity off eastern North America: its relationship to sea ice variability and climate change. Journal of Climate 7:1335–1351.

Matthews, D. R. 1993. Controlling common property: regulating Canada's east coast fisheries. University of Toronto Press, Toronto.

Mayo, R. K. 1987. Recent exploitation patterns and future stock rebuilding strategies for Acadian redfish, *Sebastes fasciatus* Storer, in the Gulf of Maine–Georges Bank region of the northwest Atlantic. Pages 334–353 in Proceedings of the international rockfish symposium. University of Alaska Sea Grant, Report 87-2, Fairbanks.

Mayo, R. K., M. J. Fogarty, and F. M. Serchuk. 1992. Aggregate fish biomass and yield on Georges Bank, 1960–87. Journal of Northwest Atlantic Fishery Science 14:59–78.

Mayo, R. K., J. M. McGlade, and S. H. Clark. 1989. Patterns of exploitation and biological status of pollock *Pollachius virens* L. in the Scotian Shelf, Gulf of Maine, and Georges Bank area. Journal of Northwest Atlantic Fishery Science 9:13–26.

McCay, B. J. 1996a. Foxes and others in the henhouse: environmentalists and the fishing industry in the U.S. regional council system. Pages 380–390 in R. M. Meyer and five coeditors. Fisheries resource utilization and policy. Proceedings of the World Fisheries Congress, Theme 2. Oxford & IBH, New Delhi.

McCay, B. J. 1996b. Participation of fishers in fisheries management. Pages 60–75 in R. M. Meyer and five coeditors. Fisheries resource utilization and policy. Proceedings of the World Fisheries Congress, Theme 2. Oxford & IBH, New Delhi.

McCay, B. J., and J. A. Acheson. 1987. The question of the commons: the culture and ecology of communal resources. University of Arizona Press, Tucson.

McCay, B. J., R. Apostle, and C. F. Creed. In press. Individual transferable quotas, comanagement, and community: lessons from Nova Scotia. Fisheries 23.

McCay, B. J., R. Apostle, C. F. Creed, A. C. Finlayson, and K. Mikalsen. 1995. Individual transferable quotas (ITQs) in Canadian and US fisheries. Ocean and Coastal Management 28:85–116.

McGarvey, R., and J. H. M. Willison. 1995. Rationale for a marine protected area along the international boundary between U.S. and Canadian waters in the Gulf of Maine. Pages 74–81 in Shackell and Willison (1995).

McGoodwin, J. R. 1990. Crisis in the world's fisheries. Stanford University Press, Stanford, California.

McKean, M. A. 1992. Success on the commons: a comparative examination of institutions for common property resource management. Journal of Theoretical Politics 4:247–281.

Millar, R. B., and R. A. Myers. 1990. Modelling environmentally induced change in growth for Atlantic Canada cod stocks. International Council for the Exploration of the Sea C.M. 1990/G:24, Copenhagen.

Mohn, R., and W. D. Bowen. 1996. Grey seal predation on the eastern Scotian Shelf: modelling the impact on Atlantic cod. Canadian Journal of Fisheries and Aquatic Sciences 53:2722–2738.

Morgan, M. J. 1992. Low-temperature tolerance of American plaice in relation to declines in abundance. Transactions of the American Fisheries Society 121:399–402.

Morgan, M. J., C. A. Bishop, and J. W. Baird. 1994. Temporal and spatial variation in age and length at maturity in cod in divisions 2J and 3KL. Northwest Atlantic Fisheries Organization, Scientific Council Studies 21:83–90.

Morisawa, M., K. Short, and T. Yamamoto. 1992. Legal framework for fisheries management in Japan. Pages 29–42 *in* T. Yamamoto and K. Short, editors. International perspectives on fisheries management. Zengyoren, Tokyo.

Mountain, D. G. 1991. The volume of shelf water in the Middle Atlantic Bight: seasonal and interannual variability, 1977–1987. Continental Shelf Research 11:251–267.

Mountain, D. G., and S. A. Murawski. 1992. Variation in the distribution of fish stocks on the northeast continental shelf in relation to their environment, 1980–1989. ICES Marine Science Symposium 195:424–432.

Murawski, S. A. 1991. Can we manage our multispecies fisheries? Fisheries 16(5):5–13.

Murawski, S. A. 1993a. Climate change and marine fish distributions: forecasting from historical analogy. Transactions of the American Fisheries Society 122:647–658.

Murawski, S. A. 1993b. Factors influencing by-catch and discard rates: analyses from multispecies/multifishery sea sampling. Northwest Atlantic Fisheries Organization, Scientific Council Research Document 93/115, Dartmouth, Nova Scotia.

Murawski, S. A., and J. S. Idoine. 1992. Multispecies size composition: a conservative property of exploited fishery systems? Journal of Northwest Atlantic Fishery Science 14:79–85.

Murawski, S. A., A. M. Lange, and J. S. Idoine. 1991. An analysis of technological interactions among Gulf of Maine mixed-species fisheries. ICES Marine Science Symposium 193:237–252.

Myers, R. A., S. A. Akenhead, and K. F. Drinkwater. 1988. The North Atlantic Oscillation and the ocean climate of the Newfoundland Shelf. Northwest Atlantic Fisheries Organization, Scientific Council Research Document 88/65, Dartmouth, Nova Scotia.

Myers, R. A., S. A. Akenhead, and K. Drinkwater. 1990. The influence of Hudson Bay runoff and ice-melt on the salinity of the inner Newfoundland Shelf. Atmosphere-Ocean 28:241–256.

Myers, R. A., N. J. Barrowman, J. A. Hutchings, and A. A. Rosenberg. 1995. Population dynamics of exploited fish stocks at low population levels. Science 269:1106–1108.

Myers, R. A., and N. Cadigan. 1994. Was an increase in natural mortality responsible for the collapse of northern cod? Northwest Atlantic Fisheries Organization, Scientific Council Research Document 94/52, Dartmouth, Nova Scotia.

Myers, R. A., and K. F. Drinkwater. 1989. The influence of Gulf Stream warm core rings on recruitment of fish in the northwest Atlantic. Journal of Marine Research 47:635–656.

Myers, R. A., K. F. Drinkwater, N. J. Barrowman, and J. W. Baird. 1993. Salinity and recruitment of Atlantic cod (*Gadus morhua*) in the Newfoundland region. Canadian Journal of Fisheries and Aquatic Sciences 50:1599–1609.

Myers, R. A., J. Helbig, and D. Holland. 1989. Seasonal and interannual variability of the Labrador Current and West Greenland Current. International Council for the Exploration of the Sea, C.M. 1989/C:16, Copenhagen.

Myers, R. A., J. A. Hutchings, and N. J. Barrowman. 1997. Why do fish stocks collapse? The example of cod in Atlantic Canada. Ecological Applications 7:91–106.

NAFO (Northwest Atlantic Fisheries Organization). 1995. Report of Scientific Council, 7–21 June 1995 meeting. NAFO, Scientific Council Summary Document 95/19, Dartmouth, Nova Scotia.

Nagasaki, F., and S. Chikuni. 1989. Management of multispecies resources and multigear fisheries. FAO (Food and Agriculture Organization of the United Nations) Fisheries Technical Paper 305.

Nakashima, B. S. 1996. The relationship between oceanographic conditions in the 1990s and changes in spawning behaviour, growth and early life history of capelin (*Mallotus villosus*). Northwest Atlantic Fisheries Organization, Scientific Council Studies 24:55–68.

Native Council of Nova Scotia. 1994. Mi'kmaq fisheries Netukulimk: towards a better understanding. Native Council of Nova Scotia, Truro.

NEFMC (New England Fishery Management Council). 1977. Final environmental impact statement for the implementation of a fishery management plan for Atlantic groundfish. NEFMC, Saugas, Massachusetts.

NEFMC (New England Fishery Management Council). 1985. Fishery management plan for the northeast multi-species fishery. NEFMC, Saugus, Massachusetts.

NEFMC (New England Fishery Management Council). 1993. Final amendment #5 to the northeast multispecies fishery management plan incorporating the supplemental environmental impact statement, volume 1. NEFMC, Saugas, Massachusetts.

NEFSC (Northeast Fisheries Science Center). 1987. Status of mixed species demersal finfish resources in New England and scientific basis for management. National Marine Fisheries Service, Woods Hole Laboratory, Reference Document 87-08, Woods Hole, Massachusetts.

NEFSC (Northeast Fisheries Science Center). 1991. Report of the 12th NEFC stock assessment workshop, spring 1991. National Marine Fisheries Service, Woods Hole Laboratory, Reference Document 91-03, Woods Hole, Massachusetts.

NEFSC (Northeast Fisheries Science Center). 1995. Status of fishery resources off the northeastern United States for 1994. NOAA (National Oceanic and Atmospheric Administration) Technical Memorandum NMFS-NE-108, Woods Hole, Massachusetts.

Neher, P. A., R. Arnason, and N. Mollett, editors. 1988. Rights based fishing. Kluwer, Dordrecht.

Nielsen, L. A., and V. G. Wespestad. 1994. The organization and implementation of the World Fisheries Congress. Pages 6–18 *in* C. W. Voigtlander, editor. The state of the world's fisheries resources. Proceedings of the World Fisheries Congress, plenary sessions. Oxford & IBH, New Delhi.

NRCUS (National Research Council of the United States) and RSC (Royal Society of Canada). 1985. The Great Lakes water quality agreement: an evolving instrument for ecosystem management. National Academy Press, Washington, D.C.

Nunallee, E. P. 1991. An investigation of the avoidance reactions of Pacific whiting (*Merluccius productus*) to demersal and midwater trawl gear. International Council for the Exploration of the Sea, Fish Capture Committee, C.M. 1990/B:5, Copenhagen.

O'Boyle, R. N., and K. C. T. Zwanenburg. 1994. Report of the Scotia-Fundy regional advisory process (RAP). Canadian Manuscript Report of Fisheries and Aquatic Sciences 2252.

O'Brien, L., J. Burnett, and R. K. Mayo. 1993. Maturation of nineteen species of finfish off the northeastern coast of the United States, 1985–1990. NOAA (National Oceanic and Atmospheric Administration) Technical Report NMFS (National Marine Fisheries Service) 113.

Ommer, R. E. 1995. Fisheries policy and the survival of fishing communities in eastern Canada. Pages 307–322 *in* A. G. Hopper, editor. Deep-water fisheries of the north Atlantic oceanic slope. Kluwer, Dordrecht, The Netherlands.

Ostrom, E. 1990. Governing the commons: the evolution of institutions for collective action. Cambridge University Press, Cambridge, UK.

Overholtz, W. J., S. A. Murawski, and K. L. Foster. 1991. Impact of predatory fish, marine mammals and seabirds on the pelagic fish ecosystem of the northeastern USA. ICES Marine Science Symposium 193:198–208.

Overholtz, W. J., and A. V. Tyler. 1985. Long-term responses of the demersal fish assemblages of Georges Bank. U.S. National Marine Fisheries Service Fishery Bulletin 83:507–520.

Overholtz, W. J., and A. V. Tyler. 1986. An exploratory simulation model of competition and predation in a demersal fish assemblage on Georges Bank. Transactions of the American Fisheries Society 115:805–817.

Parsons, L. S. 1993. Management of marine fisheries in Canada. Canadian Bulletin of Fisheries and Aquatic Sciences 225.

Pearson, J. C. 1972. The fish and fisheries of colonial North America. Pages 245–255 *in* J. C. Pearson, editor. A documentary history of fishery resources of the United States, part II. The New England States. National Technical Information Service, Springfield, Virginia.

Pella, J. J., and P. K. Tomlinson. 1969. A generalized stock production model. Inter-American Tropical Tuna Commission Bulletin 13:421–458.

Perciasepe, R. 1993. EPA's watershed approach to ecosystem management. Fisheries 19(4):4.

Perry, I. A., and S. J. Smith. 1994. Identifying habitat associations of marine fishes using survey data: an application to the northwest Atlantic. Canadian Journal of Fisheries and Aquatic Sciences 51:589–602.

Petrie, B., and C. Anderson. 1983. Circulation on the Newfoundland continental shelf. Atmosphere-Ocean 21:207–226.

Petrie, B., and K. Drinkwater. 1993. Temperature and salinity variability on the Scotian Shelf and in the Gulf of Maine 1945–1990. Journal of Geophysical Research 98:20079–20089.

Petrie, B., K. Drinkwater, and P. Yeats. 1994. Ocean climate variations for the Canadian east coast: a simple model with an update for 1993. Department of Fisheries and Oceans, Atlantic Fisheries Research Document 94/17, Dartmouth, Nova Scotia.

Petrie, B., J. W. Loder, S. Akenhead, and J. Lazier. 1992. Temperature and salinity variability on the eastern Newfoundland Shelf: the residual field. Atmosphere-Ocean 30:120–139.

Pierce, D. E. 1982. Development and evolution of fishery management plans for cod, haddock and yellowtail flounder. Massachusetts Division of Marine Fisheries, Publication 13233-133-50-5-83-CR, Boston.

Pinhorn, A. T., and R. G. Halliday. 1990. Canadian versus international regulation of northwest Atlantic fisheries: management practices, fishery yields, and resource trends, 1960–1986. North American Journal of Fisheries Management 10:154–174.

Pinkerton, E., editor. 1989. Co-operative management of local fisheries: new directions for improved management and community development. University of British Columbia Press, Vancouver.

Pinkerton, E. 1994. Local fisheries co-management: a review of international experiences and their implications for salmon management in British Columbia. Canadian Journal of Fisheries and Aquatic Sciences 51:1–17.

Pinkerton, E., and M. Weinstein. 1995. Fisheries that work: sustainability through community-based management. The David Suzuki Foundation, Vancouver.

Plan Development Team. 1990. The potential of marine fishery reserves for reef fish management in the U.S. southern Atlantic. NOAA (National Oceanic and Atmospheric Administration) Technical Memorandum NMFS-SEFC-261, Miami.

Polacheck, T. 1990. Year around closed areas as a management tool. Natural Resource Modeling 4:327–354.

Polacheck, T., D. Mountain, D. McMillan, W. Smith, and P. Berrien. 1992. Recruitment of the 1987 year class of Georges Bank haddock (*Melanogrammus aeglefinus*): the influence of unusual larval transport. Canadian Journal of Fisheries and Aquatic Sciences 49:484–496.

Policansky, D. 1993. Fishing as a cause of evolution in fishes. Pages 2–18 *in* T. K. Stokes, J. M. McGlade, and R. Law, editors. The exploitation of evolving resources. Lecture Notes in Biomathematics 99. Springer-Verlag, Berlin.

Pomeroy, R. S., and M. J. Williams. 1994. Fisheries co-management and small-scale fisheries: a policy brief. International Center for Living Aquatic Resources Management, Manila.

Pope, J. G. 1991. The ICES Multispecies Assessment Working Group: evolution, insights, and future problems. ICES Marine Science Symposium 193:22–33.

Prinsenberg, S., and I. Peterson. 1994. Interannual variability in atmospheric and ice cover properties along Canada's east coast for 1962 to 1992. Pages 372–381 *in* Proceedings of the 12th international symposium on ice, volume 1. Norwegian Institute of Technology, Trondheim.

Quinn, J. F., S. R. Wing, and L. W. Botsford. 1993. Harvest refugia in marine invertebrate fisheries: models and applications to the red sea urchin, *Strongylocentrotus franciscanus*. American Zoologist 33:537–550.

Rago, P. J., K. Sosebee, J. Brodziak, and E. D. Anderson. 1994. Distribution and dynamics of northwest Atlantic spiny dogfish (*Squalus acanthias*). National Marine Fisheries Service, Northeast Fisheries Science Center, Reference Document 94-19, Woods Hole, Massachusetts.

Raiffa, H. 1970. Decision analysis: introductory lectures on choices under uncertainty. Addison-Wesley, Reading, Massachusetts.

Rice, M. A., C. Hickox, and I. Zehra. 1989. Effects of intensive fishing effort on the population structure of quahogs, *Mercenaria mercenaria* (Linnaeus, 1758), in Narragansett Bay. Journal of Shellfish Research 8:345–354.

Ricker, W. E. 1954. Stock and recruitment. Journal of the Fisheries Research Board of Canada 11:559–623.

Ricker, W. E. 1981. Changes in the average size and average age of Pacific salmon. Canadian Journal of Fisheries and Aquatic Sciences 38:1636–1656.

Rijnsdorp, A. D., N. Daan, F. A. van Beek, and H. J. L. Heessen. 1991. Reproductive variability in North Sea plaice, sole and cod. Journal du Conseil, Conseil International pour l'Exploration de la Mer 47:253–375.

Rijnsdorp, A. D., and F. A. van Beek. 1991. The effects of the plaice box on the reduction in discarding and on the level of recruitment of North Sea sole. International Council for the Exploration of the Sea C.M. 1991/G:47, Copenhagen.

Rijnsdorp, A. D. 1993. Fisheries as a large-scale experiment on life-history evolution: disentangling phenotypic and genetic effects in changes in maturation and reproduction of North Sea plaice, *Pleuronectes platessa* L. Oecologia 96:391–401.

Rivard, D., and J.-J. Maguire. 1993. Reference points for fisheries management: the eastern Canadian experience. Pages 31–57 *in* Smith et al. (1993).

Roberts, C. M., and N. V. C. Polunin. 1991. Are marine reserves effective in management of reef fisheries? Reviews in Fish Biology and Fisheries 1:65–91.

Robertson, J. H. B. 1988. Square and diamond mesh in trawl and seine net cod-end selectivity. Pages 7–15 *in* E. Richardson, editor. Proceedings of the stock conservation engineering workshop. Rhode Island Sea Grant, Kingston.

Robertson, J. H. B., and R. S. T. Ferro. 1988. Mesh selection within the cod-ends of trawls. The effects of narrowing the cod-end and shortening the extension. Scottish Fisheries Research Report 39, Marine Laboratory, Aberdeen, Scotland.

Robins, C. R., and six coauthors. 1991a. Common and scientific names of fishes from the United States and Canada, 5th edition. American Fisheries Society Special Publication 20, Bethesda, Maryland.

Robins, C. R., and six coauthors. 1991b. World fishes important to North Americans. American Fisheries Society Special Publication 21, Bethesda, Maryland.

Rogers, J. C. 1984. The association between the North Atlantic Oscillation and the Southern Oscillation in the northern hemisphere. Monthly Weather Review 112:1999–2015.

Rose, G. A. 1993. Cod spawning on a migration highway in the north-west Atlantic. Nature (London) 366:458–461.

Rose, G. A., B. A. Atkinson, J. Baird, C. A. Bishop, and D. W. Kulka. 1994. Changes in distribution of Atlantic cod and thermal variations in Newfoundland waters, 1980–1992. ICES Marine Science Symposium 198:542–552.

Rosenberg, A. A., M. J. Fogarty, M. P. Sissenwine, J. R. Beddington, and J. G. Shepard. 1993. Achieving sustainable use of renewable resources. Science 262:828–829.

Rothschild, B. J. 1973. Questions of strategy in fishery management and development. Journal of the Fisheries Research Board of Canada 30:2017–2030.

Rothschild, B. J., and A. J. Mullen. 1985. The information content of stock and recruitment data and its non-parametric classification. Journal du Conseil, Conseil International pour l'Exploration de la Mer 42:116–126.

Ruddle, K. 1994. A guide to the literature on traditional community-based fishery management in the Asia–Pacific tropics. FAO Fisheries Circular 869.

Ruddle, K., E. Hviding, and R. E. Johannes. 1992. Marine resources management in the context of customary tenure. Marine Resource Economics 7:249–273.

Ryman, N., and F. Utter, editors. 1987. Population genetics and fishery management. University of Washington Press, Seattle.

SAFMC (South Atlantic Fishery Management Council). 1993. Amendment 6, regulatory impact review, initial regulatory flexibility analysis and environmental assessment for the snapper grouper fishery of the south Atlantic region. SAFMC, Charleston, South Carolina.

Schaefer, M. B. 1954. Some aspects of the dynamics of populations important to the management of the commercial marine fisheries. Inter-American Tropical Tuna Commission Bulletin 1:27–56.

Scott, A. 1993. Obstacles to fishery self-government. Marine Resource Economics 8:187–199.

Scott, J. S. 1982. Depth, temperature and salinity preferences of common fishes of the Scotian Shelf. Journal of Northwest Atlantic Fishery Science 3:29–39.

Serchuk, F. M., M. D. Grosslein, R. G. Lough, D. G. Mountain, and L. O'Brien. 1994a. Fishery and environmental factors affecting trends and fluctuations in the Georges Bank and Gulf of Maine Atlantic cod stocks: an overview. ICES Marine Science Symposium 198:77–109.

Serchuk, F. M., R. K. Mayo, and L. O'Brien. 1994b. Assessment of the Georges Bank cod stock for 1994. National Marine Fisheries Service, Northeast Fisheries Science Center, Reference Document 94-25, Woods Hole, Massachusetts.

Serchuk, F. M., and R. J. Solowitz. 1990. Ensuring fisheries management dysfunction: the neglect of science and technology. Fisheries 15(2):4–7.

Serchuk, F. M., and S. E. Wigley. 1992. Assessment and management of the Georges Bank cod fishery: an historical review and evaluation. Journal of Northwest Atlantic Fishery Science 13:25–52.

Shackell, N. L., and J. H. M. Willison, editors. 1995. Marine protected areas and sustainable fisheries. Science and Management of Protected Areas Association, Wolfville, Nova Scotia.

Shelton, P. A. 1995. Analysis of replacement in eight northwest Atlantic cod stocks. Department of Fisheries and Oceans, Atlantic Fisheries Research Document 95/75, Dartmouth, Nova Scotia.

Shelton, P. A., and D. B. Atkinson. 1994. Failure of the div. 2J3KL cod recruitment prediction using salinity. Department of Fisheries and Oceans, Atlantic Fisheries Research Document 94/66, Dartmouth, Nova Scotia.

Shelton, P. A., G. R. Lilly, and E. Colbourne. 1996. Patterns in the annual weight increment for 2J3KL cod and possible prediction for stock projection. Northwest Atlantic Fisheries Organization, Scientific Council Research Document 96/47, Dartmouth, Nova Scotia.

Shelton, P. A., and M. J. Morgan. 1994. NAFO divisions 3NO cod stock—spawner stock biomass and recruitment required for replacement. Northwest Atlantic Fisheries Organization, Scientific Council Research Document 94/1, Dartmouth, Nova Scotia.

Shelton, P. A., G. B. Stenson, B. Sjare, and W. G. Warren. 1995. Model estimates of harp seal numbers at age for the northwest Atlantic. Department of Fisheries and Oceans, Atlantic Fisheries Research Document 95/21, Dartmouth, Nova Scotia.

Shepherd, J. G. 1982. A versatile new stock-recruitment relationship of fisheries and construction of sustainable yield curves. Journal du Conseil, Conseil International pour l'Exploration de la Mer 40:87–75.

Shepherd, J. G., J. G. Pope, and R. D. Cousens. 1984. Variations in fish stocks and hypotheses concerning their links with climate. Rapports et Procès-Verbaux des Réunions, Conseil International pour l'Exploration de la Mer 185:255–267.

Sherman, K. L. 1992. Monitoring and assessment of large marine ecosystems: a global and regional perspective. Pages 1041–1074 in D. H. Mckenzie, D. E. Hyatt, and V. J. McDonald, editors. Ecological indicators, volume 2. Elsevier, Lancaster, UK.

Sherman, K. L. 1994. Sustainability, biomass yields, and health of coastal ecosystems: an ecological perspective. Marine Ecology Progress Series 112:277–301.

Simmel, G. 1955. Conflict and the web of group affiliations. Free Press, New York.

Sinclair, A. 1996. Recent declines in cod species stocks in the northwest Atlantic. Northwest Atlantic Fisheries Organization, Scientific Council Studies 24:41–52.

Sinclair, A., D. Gascon, R. O'Boyle, D. Rivard, and S. Gavaris. 1991. Consistency of some northwest Atlantic groundfish stock assessments. Northwest Atlantic Fisheries Organization, Scientific Council Studies 16:59–77.

Sinclair, A. F., and seven coauthors. 1994. Assessment of the fishery for southern Gulf of St. Lawrence cod: May 1994. Department of Fisheries and Oceans, Atlantic Fisheries Research Document 94/77, Dartmouth, Nova Scotia.

Sinclair, A. F., and seven coauthors. 1995. Assessment of the southern Gulf of St. Lawrence cod stock, March 1995. Department of Fisheries and Oceans, Atlantic Fisheries Research Document 95/39, Dartmouth, Nova Scotia.

Sinclair, W. F. 1978. Management alternatives and strategic planning for Canada's fisheries. Journal of the Fisheries Research Board of Canada 35:1017–1030.

Sissenwine, M. P., and A. A. Rosenberg. 1993. Marine fisheries at a critical juncture. Fisheries 18(10):6–14.

Sissenwine, M. P., and J. G. Shepherd. 1987. An alternative perspective on recruitment overfishing and biological reference points. Canadian Journal of Fisheries and Aquatic Sciences 44:913–918.

Slocombe, S. 1993. Implementing ecosystem-based management. Development of theory, practice, and research for planning and managing a region. BioScience 43:612–622.
Smith, L. J., and T. R. Bacek. 1992. Fish facts and fables. SOS Publications, Fairhaven, New Jersey.
Smith, P. C., and F. B. Schwing. 1991. Mean circulation and variability on the eastern Canadian continental shelf. Continental Shelf Research 11:977–1012.
Smith, P. J., R. I. C. C. Francis, and M. McVeagh. 1991. Loss of genetic diversity due to fishing pressure. Fisheries Research (Amsterdam) 10:309–316.
Smith, S. J., J. J. Hunt, and D. Rivard, editors. 1993. Risk evaluation and biological reference points for fisheries management. Canadian Special Publication of Fisheries and Aquatic Sciences 120.
Smith, S. J., and F. Page. 1996. Associations between Atlantic cod (*Gadus morhua*) and hydrographic variables: implications for the management of the 4VsW cod stock. ICES Journal of Marine Science 53:597–614.
Smith, S. J., R. I. Perry, and L. P. Fanning. 1991. Relationships between water mass characteristics and estimates of fish population abundance from trawl surveys. Environmental Monitoring and Assessment 17:227–245.
Smith, T. D. 1994. Scaling fisheries: the science of measuring the effects of fishing, 1855–1955. Cambridge University Press, New York.
Staatz, J. M. 1989. Farmer cooperative theory: recent developments. U.S. Department of Agriculture, ACS (Agricultural Cooperative Service) Research Report 84.
Stein, M. 1993. On the consistency of thermal events in the East Greenland/West Greenland current system and off Labrador. Northwest Atlantic Fisheries Organization, Scientific Council Studies 14:29–37.
Stenson, G. B., M. O. Hammill, and J. W. Lawson. 1995. Predation of Atlantic cod, capelin, and Arctic cod by harp seals in Atlantic Canada. Department of Fisheries and Oceans, Atlantic Fisheries Research Document 95/72, Dartmouth, Nova Scotia.
Stephenson, R. L., and D. E. Lane. 1995. Fisheries management science: a plea for conceptual change. Canadian Journal of Fisheries and Aquatic Sciences 52:2051–2056.
Sutcliffe, W. H., Jr. 1972. Some relations of land drainage, nutrients, particulate material, and fish catch in two Canadian bays. Journal of the Fisheries Research Board of Canada 29:357–362.
Sutcliffe, W. H., Jr. 1973. Correlations between seasonal river discharge and local landings of American lobster (*Homarus americanus*) and Atlantic halibut (*Hippoglossus hippoglossus*) in the Gulf of St. Lawrence. Journal of the Fisheries Research Board of Canada 30:856–859.
Sutcliffe, W. H., Jr., K. Drinkwater, and B. S. Muir. 1977. Correlations of fish catch and environmental factors in the Gulf of Maine. Journal of the Fisheries Research Board of Canada 34:19–30.
Sutcliffe, W. H., Jr., R. H. Loucks, and K. F. Drinkwater. 1976. Coastal circulation and physical oceanography of the Scotian Shelf and the Gulf of Maine. Journal of the Fisheries Research Board of Canada 33:98–115.
Sutcliffe, W. H., Jr., R. H. Loucks, K. F. Drinkwater, and A. R. Coote. 1983. Nutrient flux onto the Labrador Shelf from Hudson Strait and its biological consequences. Canadian Journal of Fisheries and Aquatic Sciences 40:1692–1701.
Sutinen, J. G., A. Rieser, and J. R. Gauvin. 1990. Measuring and explaining non-compliance in federally managed fisheries. Ocean Development and International Law 21:335–372.
Taggart, C. T., and 10 coauthors. 1994. Overview of cod stocks, biology, and environment in the northwest Atlantic region of Newfoundland, with emphasis on northern cod. ICES Marine Science Symposium 198:140–157.
Templeman, W., and J. Gulland. 1965. Review of possible conservation actions for the ICNAF area. International Commission for the Northwest Atlantic Fisheries Annual Proceedings 15:47–56.

TFIAAF (Task Force on Incomes and Adjustment in the Atlantic Fishery). 1993. Charting a new course: towards the fishery of the future. Department of Fisheries and Oceans, Miscellaneous Publication Fs 23-243/1993, Ottawa.

Thompson, K. R., and F. H. Page. 1989. Detecting synchrony of recruitment using short, autocorrelated time series. Canadian Journal of Fisheries and Aquatic Sciences 46:1831–1838.

Townsend, R. E. 1990. Entry restrictions in the fishery: a survey of the evidence. Land Economics 66:359–378.

Townsend, R. E. 1992. A fractional licensing program for fisheries. Land Economics 68:185–190.

Trippel, E. A. 1995. Age at maturity as a stress indicator in fisheries. BioScience 45:759–771.

Turgeon, D. D., and nine coauthors. 1988. Common and scientific names of aquatic invertebrates from the United States and Canada: mollusks. American Fisheries Society Special Publication 16, Bethesda, Maryland.

Tyler, A. V. 1972. Food resource division among northern, marine, demersal fishes. Journal of the Fisheries Research Board of Canada 29:997–1003.

Umoh, J. U. 1992. Seasonal and interannual variability of sea temperature and surface heat fluxes in the northwest Atlantic. Doctoral dissertation. Dalhousie University, Halifax, Nova Scotia.

Umoh, J. U., J. W. Loder, and B. Petrie. 1995. The role of air–sea heat fluxes in annual and interannual ocean temperature variability on the eastern Newfoundland Shelf. Atmosphere-Ocean 33:531–568.

Wahle, R. A., and R. S. Steneck. 1991. Recruitment habitats and nursery grounds of American lobster (*Homarus americanus* Milne Edwards): a demographic bottleneck. Marine Ecology Progress Series 69:231–243.

Wahle, R. A., and R. S. Steneck. 1992. Habitat restrictions in early benthic life: experiments on habitat selection and in situ predation with American lobster. Journal of Experimental Marine Biology and Ecology 157:91–114.

Walsh, S. J. 1992. Factors influencing distribution of juvenile yellowtail flounder (*Limanda ferruginea*) on the Grand Bank of Newfoundland. Netherlands Journal of Sea Research 29:193–203.

Walsh, S. J., W. B. Brodie, C. Bishop, and E. Murphy. 1995. Fishing on juvenile groundfish nurseries on the Grand Bank: a discussion of technical measures of conservation. Pages 54–73 *in* Shackell and Willison (1995).

Walters, C. 1986. Adaptive management of renewable resources. Macmillan, New York.

Walters, C., R. D. Goruk, and D. Radford. 1993. Rivers Inlet sockeye salmon: an experiment in adaptive management. North American Journal of Fisheries Management 13:253–262.

Wardle, C. S. 1986. Fish behavior and fishing gear. Pages 463–495 *in* T. J. Pitcher, editor. The behavior of teleost fishes. Croom Helm, London.

Weare, B. C. 1977. Empirical orthogonal analysis of Atlantic Ocean surface temperatures. Quarterly Journal of the Royal Meteorological Society 103:467–478.

Wells, R., and A. T. Pinhorn. 1974. Some implications of closure of Hamilton Inlet Bank to the commercial cod fishery during the spawning season. International Commission for the Northwest Atlantic Fisheries, Research Document 74/104, Dartmouth, Nova Scotia.

Werner, F. E., and seven coauthors. 1993. Influences of mean advection and simple behavior on the distribution of cod and haddock early life stages on Georges Bank. Fisheries Oceanography 2:43–64.

Williams, A. B., and six coauthors. 1989. Common and scientific names of aquatic invertebrates from the United States and Canada: decapod crustaceans. American Fisheries Society Special Publication 17, Bethesda, Maryland.

Wilson, E. O. 1984. Biophilia. Harvard University Press, Cambridge, Massachusetts.
Wilson, J. A., J. M. Acheson, M. Metcalfe, and P. Kleban. 1994. Chaos, complexity and community management of fisheries. Marine Policy 18:291–305.
Wilson, J. A., and P. Kleban. 1991. Practical implications of chaos in fisheries: ecologically adapted management. Maritime Anthropological Studies 5:67–75.
Wilson, J. A., R. Townsend, P. Kleban, S. McKay, and J. French. 1990. Managing unpredictable resources: traditional policies applied to chaotic populations. Ocean & Shoreline Management 13:179–197.
Wooster, W. S., editor. 1988. Fishery science and management: objectives and limitations. Lecture notes on coastal and estuarine studies, volume 28. Springer-Verlag, New York.
Young, M. D. 1992. Sustainable investment and resource use. Parthenon Press, Carnforth, UK.

GLOSSARY

Abundance: The relative number of individuals (e.g., fish) in a defined group.

Advection: Horizontal movement of water (or air) and the organisms (e.g., fish eggs and larvae), other particles, and chemicals carried in the water.

Age at maturity: The median or average age at which fish in a stock first become capable of reproduction.

Age-class (age-group): A group of fish with the same chronological age (e.g., age-3 haddock).

Beam trawl: A bottom trawl whose mouth is held open by a rigid frame.

Biomass: The total weight of organisms in a defined group, such as a fish stock or year-class.

Broadcast spawners: Fish and other animals that discharge their eggs and sperm into the water.

Bycatch: The incidental catch of species other than the one(s) targeted.

Catch per unit effort: The number or weight of fish (or other species) caught divided by the amount of fishing effort expended to catch them (e.g., tons per day of fishing).

Cohort: A group of fish in a stock spawned at the same time, often referring to a particular year-class.

Correlation analysis: A statistical method to determine how faithfully one factor varies in relation to another.

Customary marine tenure: A right to fish established by long community tradition.

Demersal: Living on or near the bottom of the ocean.

Directed fishery: A fishery that targets a particular species, a particular group of species, or a particular subgroup of a species (e.g., large cod, male lobsters).

Discard: A discarded fish. See *discarding*.

Discarding: The abandonment, usually at sea, of unwanted or illegal fish brought up in the catch.

Distant-water fleet: Vessels fishing far from their home ports; specifically, non-North American vessels fishing in or near U.S. or Canadian waters.

Dragger: A bottom trawler; a person who fishes the bottom with an otter trawl.

F: The symbol for instantaneous fishing mortality.

$F_{0.1}$: The fishing mortality at which an additional unit of fishing effort brings 10% of the added yield that it would bring in a virgin fishery.

F_{MAX}: The fishing mortality that maximizes yield per recruit, as determined by dynamic pool models.

F_{MSY}: The fishing mortality that generates maximum sustainable yield, as determined by surplus production models.

Fecundity: The number of eggs produced by a female.

Fishing effort: The aggregate amount of time spent and gear used in a fishery over a designated period such as a season or year. Units of fishing effort vary among fisheries; examples are trap night (one trap fished for one night), vessel day, and trawling hour.

Fishing mortality: The rate of death in a fish stock as a result of fish harvesting, denoted F by convention. As used in fishery models, it is an "instantaneous" rate that functions like negative compound interest.

Fixed gear: Fishing gear that is set in one place until it is retrieved (e.g., gill nets, traps, longlines). See *mobile gear*.

Genetic diversity: The amount of genetic variation within a species or stock. It is believed that greater genetic diversity makes a stock better able to withstand stresses imposed by a fishery or environmental change. Harvests that disproportionately remove particular subgroups of fish, such as fast-growing fish or those adapted to local conditions, will reduce genetic diversity in a stock.

Growth overfishing: Excessive harvest of fish before they have had a chance to grow to a size that would produce maximum yield. See *recruitment overfishing*.

Gyre: A circular motion or form, such as the rotating water masses that break off the Gulf Stream and other currents.

High-grading: Selective discarding of poor quality and unwanted fish at sea to increase the value of the landed catch.

ICNAF: International Commission for the Northwest Atlantic Fisheries.

Individual transferable quota: A fixed portion of the total allowable catch that can be bought and sold by a license holder.

ITQ: Individual transferable quota.

Latent heat flux: Gain or loss of heat without change in temperature, as by evaporation or condensation. See *sensible heat loss*.

Limited entry: A restriction on the number of people or vessels allowed to participate in a fishery.

M: The symbol for instantaneous natural mortality.

Maximum sustainable (sustained) **yield**: The theoretical maximum weight of fish that could be harvested from a stock year after year if the stock reacted to fishing according to surplus production models.

Mobile gear: Fishing gear that moves through the water as it catches fish (e.g., trawls, seines). See *fixed gear*.

Mortality: The rate of death. See *fishing mortality, natural mortality, total mortality*.

MSY: Maximum sustainable yield.

NAFO: Northwest Atlantic Fisheries Organization.

NAFO convention area: The area of the Atlantic Ocean under the fisheries jurisdiction of the Northwest Atlantic Fisheries Organization.

NAO: North Atlantic Oscillation.

Natural mortality: The rate of death in a fish stock from all causes other than fishing, denoted M by convention.

Nontarget species: A species that is not the desired species of a directed fishery but that may be captured as bycatch.

North Atlantic Oscillation: Reciprocal changes in atmospheric pressure between a low-pressure system centered near Iceland (the Icelandic Low) and a high-pressure system centered over the Azores (the Azores High).

Optimum yield (optimum sustainable yield): The maximum sustainable biological yield modified to "optimize" the economic and social benefits from a fishery.

Otter trawl: A trawl whose mouth is held open by water pressure acting on boards (otter boards) affixed to the trawl's bridle. Otter trawls can be fished on the bottom or in midwater.

Pair trawling: A fishing technique whereby two boats are used to pull one bottom trawl.

Pelagic: Living in open water, above the bottom.

Practical salinity: The ratio of the mass of material dissolved in seawater to the mass of the seawater, estimated from the conductivity of a seawater sample. Multiplying practical salinity by 1,000 gives the number of "practical salinity units" (psu), which have the same numerical value as salinity determined by older titration methods and expressed as "parts per thousand."

Recruit: A fish that has become large or old enough to be captured by the fishing gear in use.

Recruitment: The growth of fish into the harvestable portion of a stock.

Recruitment overfishing: Excessive harvest of mature fish such that production of young fish becomes too low to sustain stock size. See *growth overfishing*.

Risk aversion: Use of "conservative" fishery regulations to reduce the chance that stocks will be overfished as a result of inaccurate population analyses.

Sensible heat flux: Gain or loss of heat accompanied by a change in temperature, as by solar radiation or wind convection. See *latent heat loss*.

Shelf water: Water on or above the continental shelf, the gently dipping sea bottom extending from shore.

Size-specific fishing mortality: The fishing mortality associated with the harvest of a particular size of fish.

Slope water: Water on or above the continental slope, the steeply dipping sea bottom extending from the continental shelf to the ocean abyss.

Spawning potential: The estimated biomass of spawners likely to be produced by a class of recruits at a designated level of fishing mortality.

Spawning potential ratio: The ratio of spawning potential in a fished stock to the potential in an unfished stock.

Spawning stock biomass: The aggregate weight of mature fish in a stock.

Spawning stock biomass per recruit: The average weight each recruit will reach after it becomes sexually mature.

Species diversity: A combined measure of the number of species present and the number of individuals per species.

Stationary gear: See *fixed gear*.

Stock: A more-or-less discrete and identifiable unit of a fish or other exploited species, often referring to a management unit.

Stock assessment: Use of fishery data, research surveys, and population models to estimate the status of a stock and the yield that can be taken from it.

Stock–recruitment relationship: The biomass or number of recruits expected to be produced by a given biomass or number of spawners.

Surplus production: The biomass of fish that can be removed from a stock without harming the stock's ability to sustain itself.

TAC: Total allowable catch.

Targeted fishery: See *directed fishery*.

Territorial use right in fishing: A traditional, community-sanctioned right to fish a particular area.

Total allowable catch: The maximum overall catch from a stock permitted by a fisheries management agency during a specified period of time, usually a year.

Total mortality: The rate of death in a fish stock from all causes, denoted Z by convention. Total mortality is the sum of fishing and natural mortalities.

Virtual population analysis: A mathematical analysis in which catch data are used to estimate absolute abundances of age-classes in a stock and the fishing mortality that has been sustained by those age-classes.

VPA: Virtual population analysis.

Warm core ring: A ring or gyre of warm water that breaks away from the Gulf Stream (or any warm current) as it moves northward.

Year-class: A group of fish hatched in the same calendar year (e.g., the 1993 year-class of Atlantic cod).

Yield: The production from a fishery in terms of number or weight of fish.

Yield per recruit: The average weight of fish ultimately harvested for each fish that enters the fishery.

Z: The symbol for instantaneous total mortality.

INDEX

age-classes 76
Alaskan Cost Recovery Salmon Enhancement Associations 190–191
assemblages, fish 82–83, 156; diagrams 148, 164
Atlantic Groundfish Advisory Committee (AGAC) 56, 57
atmospheric conditions 6–7
Azores High 6, 19

barometric pressure 6
  Azores High 6, 19
  Icelandic Low 6, 19
beam trawling 167; diagram 168
biomass 146, 171
  Atlantic cod 34, 37, 38; diagrams 34, 35, 73
  elasmobranchs 82–83, 84; diagram 83
  groundfish 73–74; diagrams 73, 83
  haddock 41
  redfish 43
  spawning 86; diagram 87
  witch flounder 47
bycatch 160, 173

Canada
  exports 53–54
  fishery management institutions 55–57, 98–99, 126–128, 129, 130, 132–134, 136, 192
  fishing industry characteristics 49, 124–125, 127, 128, 129–130, 131–132
  landings 31
  landings of witch flounder 47
  management units used by 37, 41–42, 85–86, 88, 107
  policy conflicts with France 106
  policy for inshore and offshore vessels 127–128, 130
  regulatory enforcement 105–106
  unemployment insurance 136

Canadian Atlantic groundfish management plan 56–57
Canadian Atlantic Scientific Advisory Committee (CAFSAC) 55–56
Canadian Fisheries Resource Conservation Council 192
closed areas 97–98, 99–100, 102, 159–161, 165; diagram 164
  enforcement 165
  permanent 164–165
  seasonal 162–163
Coastal Communities Network 192
Coastal Fisheries Protection Act 105
cod, Atlantic
  biomass 34, 37, 38; diagrams 35, 73, 87
  condition 77–78
  environmental factors related to 21, 22
  in Gulf of Maine and Georges Bank 37–38
  in Gulf of St. Lawrence 36–37
  juvenile survival rate 86; diagram 87
  off Labrador and Newfoundland 33–35
  landings 32–38, 62–63, 75; diagrams 34, 35, 73, 74
  management units used for Georges Bank 37
  recruitment 81; diagram 87
  on Scotian Shelf 37
  seal predation 84–85
  stock characteristics 32–38, 62–63, 76–78
cod, northern 30, 56, 57
cold intermediate layer (CIL) 13, 19; diagram 17
comanagement 123, 130, 185, 186–189, 191–193
  case studies 123, 189–191
  conditions for success 189
commercial fishing effort
  Canadian 49; diagrams 50, 51, 52
  United States 50–54; diagrams 50, 53

237

compliance   187; *see also* enforcement
computer simulation models   89–90
condition, fish   77
consolidation   133–134, 182, 183
continental shelf   3–6; map   4
corporate governance   195–201
  bylaws   196
  compared to cooperative management   197–198
  financial considerations   197
  and operating contracts   199–200
correlation analysis   79
customary marine tenure   177–178

decision-analysis process   203–204, 208; diagram   208
  constraining factors   205–206
  development and evaluation of alternatives   206–207
  management team   207–208
  objectives   206
  and problem definition   204, 205
Department of Fisheries and Oceans (DFO)   98–100, 126
  interactions of fishermen with   128, 130, 132, 133
  and snow crab management   130
discards   160, 171, 181
discount rate   197
distant-water fleets   52, 54
distribution, fish   21
diversity
  genetic   160–161
  species   82
dogfish, spiny   82–84
draggers, middle-distance   131–132
dynamic pool model   144–145; diagram   144

ecosystems   153, 156
eddies, *see* rings
eggs   22
elasmobranchs   82–84
  biomass   82–83, 84; diagram   83
enforcement   98, 100, 103, 104, 105–106, 107–108, 122
  for closed areas   165
  for individual quota programs   180
environmental factors   3, 78, 81
  atmospheric conditions   6–7
  effects on catch rates   21
  effects on fish distribution   21
  effects on fish growth   21
  effects on recruitment   22

geography   3–6
icebergs   9
river runoff   7, 79–80
salinity   12, 80
sea ice   7–9
temperature   11–12, 80–81
water circulation   9–11
ethnicity   119
European Union   104, 105–106
experimental management   154–155

F 141–142
$F_{0.1}$ 85–86, 88, 98, 104, 145; diagram   144
F_{MAX} 85–86, 144–145; diagram   144
F_{MED} 147
F_{MSY} 143, 144, 145; diagram   143
Fisheries Cooperative Associations   189–190
Fishery Conservation and Management Act, *see* Magnuson Fishery Conservation and Management Act
Fishery Management Plan for Atlantic Groundfish   101
Fishery Management Plan for the Northeast Multispecies Fishery   102–103
Fishery Resource Conservation Council (FRCC)   57
fixed gear   129–130
flatfish, *see* flounders
flounder, winter   69
flounder, witch
  biomass   47
  off Canada   47
  in the Gulf of Maine and on Georges Bank   47
  landings   47, 67
  stock characteristics   46–47, 67
flounder, yellowtail
  off Cape Cod   48
  on Georges Bank   48
  landings   47–48, 68, 75; diagram   74
  off Newfoundland   47
  off Southern New England   48
  stock characteristics   47–48, 68
flounders
  landings   30–31, 70; diagrams   32, 33
  species   32
  stock characteristics   70

Gaspé Current   9
gear   115, 167–170, 172–174; diagrams   168, 169, 171, 173
  evaluation   170
  fixed   129–130

mobile   129–130
selectivity   172–174
use rights   179–180
genetic diversity   160–161, 171
geography, as an environmental factor   3–6
Georges Bank
  American plaice stocks   46
  Atlantic cod stocks   37–38
  fishing industry characteristics   50–51, 52
  haddock stocks   40–41
  jurisdictional claims   106–107
  silver hake stocks   44
  witch flounder stocks   47
  yellowtail flounder stocks   48
gill nets   170; diagram   171
Gloucester   118–122
  and Italian fishing families   118, 120
  role of women in   118–119, 120–121
Gloucester Fishermen's Wives Association   120–121
gravel   156–157
groundfish
  biomass   73; diagrams   73, 83
  Canadian management plan for   56–57
  composition and distribution   3–4, 82; map   4
  imports   53–54; diagram   55
  landings   30, 31, 49, 50, 51, 74–75; diagrams   31, 32, 73, 74
  mortality rates   86, 88; diagrams   89, 90
  predator–prey relationships   83–85
  Scotia-Fundy fishery for   131–135
growth rates   21
Gulf of Maine
  American plaice stocks   46
  Atlantic cod stocks   37–38
  haddock stocks   40–41
  pollock stocks   45
  redfish stocks   43
  sea surface temperature   80
  silver hake stocks   43
  witch flounder stocks   47
Gulf of St. Lawrence
  air temperature   18–19
  American plaice stocks   46
  Atlantic cod stocks   36–37
  pollock stocks   45
  redfish stocks   43
  sea temperature   19; diagram   20

habitat   156–157, 161, 164–165
  and fish distribution   162; diagram   163
  and fishing gear   172
haddock
  biomass of Georges Bank stock   41
  in Gulf of Maine and Georges Bank   40–41
  landings   38–41, 64, 75; diagrams   39, 40, 74
  management units used for Georges Bank   41
  maturity rates   78; diagram   78
  off Newfoundland   38
  recruitment   81
  on Scotian Shelf   39
  stock characteristics   38–41, 64
hake, silver
  in Gulf of Maine and Northern Georges Bank   44
  landings   44, 66
  on Scotian Shelf   44
  in Southern Georges Bank and Middle Atlantic Bight   44
  stock characteristics   44, 66
halibut, Atlantic   69
  landings   75; diagram   74
halibut, Greenland
  landings   49, 68
  stock characteristics   48–49, 68
harvest rights   180–181
heat advection   7
heat exchange   7
  effects of ice on   9
herring, Atlantic   83, 84
highliner   117

ice, *see* sea ice
icebergs   9, 16
Icelandic Low   6, 15, 19
indigenous peoples   126–127, 177–178
individual quotas   99, 180
individual transferable quotas (ITQs)   128, 129–130, 131, 132–135, 180
  concentration of   133–134
  in the Scotia-Fundy groundfish fishery   134
inshore/offshore split   127–128, 130
Interim Fishery Management Plan for Atlantic Groundfish   101–102
International Commission for the Northwest Atlantic Fisheries (ICNAF)   54, 96–98, 103, 104
  enforcement   98
  management units used by   85
International Court of Justice   106–107

Japan 189–190
juvenile survival rate 86; diagram 87

Labrador
    air temperature 14–15; diagram 17
    Atlantic cod stocks 33–35
    and salinity 16, 18
    and sea ice 15–16
    sea temperature 16, 18
Labrador Current 9, 16
landings 73–74
    American plaice 46, 66–67, 75; diagram 74
    Atlantic cod 32–38, 62–63, 75; diagrams 73, 74
    Atlantic halibut 69, 75; diagram 74
    Canada 31
    flounders 30–31, 70; diagrams 32, 33
    Greenland halibut 49
    groundfish 30, 31, 74–75; diagrams 31, 33, 73, 74
    haddock 38–41, 64, 75; diagrams 39, 40, 74
    pollock 45, 66, 75; diagram 74
    redfish 41, 42–43, 65, 75; diagrams 40, 42, 74
    silver hake 44, 66
    United States 31
    winter flounder 69
    witch flounder 47, 67
    yellowtail flounder 47–48, 68, 75; diagram 74
larvae 22
latent heat flux 7
life-history characteristics 161; diagram 162
limited entry 178–179
lobster, Atlantic Canadian fishery for 179, 182–183
longlining 167; diagram 168
    effort 49; diagram 52

Magnuson Fishery Conservation and Management Act 57, 100, 101, 113–114
    enforcement 103
    national standards 101
management
    ecosystem-based 153–156, 157
    multispecies 147–150, 151
    single-species 141–142, 150–151
    units used for Georges Bank Atlantic cod 37
    units used for Georges Bank haddock 41
    units used for pollock 45

units used for redfish 41–42
maturity 78
maximum spawning potential 102
maximum sustained yield (MSY) 96
mesh size 97, 99, 101–102, 103, 104, 172
Middle Atlantic Bight
    salinity 19, 21
    sea temperature 19, 21
    silver hake stocks 44
minimum size regulations 99, 100, 102
mobile gear 129–130
monofilament 171
mortality rates 85–86, 88, 141–142, 148–149, 150; diagrams 87, 88, 89, 149
Multispecies Fishery Management Plan 116
multispecies management 147–150, 151
multispecies virtual population analysis 150

Native Peoples, see indigenous peoples
navigational technology 169–170
New England Fishery Management Council 100, 101, 102–103, 114, 116–117, 122
    interactions of fishermen with 116–117, 122, 123–124, 135
Newfoundland
    air temperature 14; diagram 16
    American plaice stocks 46
    Atlantic cod stocks 33–35
    fishing industry characteristics 49
    haddock stocks 38
    redfish stocks 42–43
    and salinity 16, 18, 80; diagram 79
    and sea ice 15–16
    sea temperature 16, 18, 80–81; diagrams 17, 79
    yellowtail flounder stocks 47
Nordmore grate 173; diagram 173
North Atlantic Oscillation (NAO) index 6, 15, 16, 19, 81; diagrams 17, 79
    five-year running mean of 17
Northeast Multispecies Fishery Management Plan 58, 151
Northwest Atlantic Fisheries Organization (NAFO) 54, 103–106
    enforcement 104–106
    management area 28, 30, 71–72; maps 5, 29, 72
    and nonmember fishing 105
Nova Scotia Current 9–10

ocean perch, see redfish

Index 241

optimum yield 101
otter trawling 27, 167–169, 173; diagram 169
  effort 52; diagrams 52, 53
  mesh-size regulations 97, 99
overcapitalization 98–99, 114, 132
overfishing 85–86, 88–89, 90–91, 121–122, 132

plaice, American
  in the Gulf of Maine and on Georges Bank 46
  in the Gulf of St. Lawrence 46
  landings 46, 66–67, 75; diagram 74
  off Newfoundland 46
  stock characteristics 45–46, 66–67
pollock
  landings 45, 66, 75; diagram 74
  management units 45
  on Scotian Shelf and in Gulf of Maine 45
  stock characteristics 45, 66
predator–prey relationships 83–85, 150, 156

recruitment 22, 81, 89–90, 145; diagram 87
redfish
  biomass 43
  in Gulf of Maine 43, 51
  in Gulf of St. Lawrence 43
  landings 41, 42–43, 65, 75; diagrams 40, 42, 74
  management units 41–42
  off Newfoundland 42–43
  on Scotian Shelf 43
  species of 42
  stock characteristics 42–43, 65
reserves, *see* closed areas
rings 11, 22
river runoff 7, 79–80

salinity 12, 16, 80; diagrams 15, 79
salmon 190–191
Scotian Shelf
  Atlantic cod stocks 37
  haddock stocks 39
  pollock stocks 45
  redfish stocks 43
  silver hake stocks 44
sea ice 7–9, 15–16; map 8
  effects on heat exchange 9
seal predation 84–85
selectivity, gear 172–174
sensible heat flux 7

single-species management 141–142, 150–151
  dynamic pool model 144–145; diagram 144
  stock–recruitment model 145–147; diagram 146
  surplus production model 143; diagram 143
slope water 11
sound, and gear selectivity 174
spawning 22, 76
spawning potential ratio 146
spawning stock biomass per recruit 145–147; diagram 147
St. Lawrence River 7
  discharge 79–80; diagram 79
stock assessment workshops 58–59
stock–recruitment model 145–147; diagram 146
stocks
  defined 141
  numbers of 30
storms 6
St. Pierre Bank 106
support industries 115–116
surplus production model 143; diagram 143

Task Force on Atlantic Fisheries 99
temperature 11–12
  air 6, 14–15, 18–19; diagrams 16, 17, 18
  effects on fish 21–22, 23
  midwater 19, 80; diagram 20
  sea 7, 16, 18, 19, 21
  sea bottom 12, 80; diagrams 14, 17, 79
  sea surface 12, 80; diagrams 13, 20, 79
territorial use rights 177–178
tidal currents 11
total allowable catch 97, 205
transboundary stocks 105–107
trawling 27, 49
  beam 167; diagram 168
  effort 52; diagrams 52, 53
  mesh size regulations 97, 99
  otter 27, 167–169, 173; diagram 169
  pair 122
turbot, *see* halibut, Greenland

uncertainty, scientific 187
unemployment insurance 136
United Nations Conference on Straddling and Highly Migratory Fish Stocks 105

United Nations Convention on the Law of
    the Sea   105
United States
    fishery management institutions   57–59,
        100–103, 113–114, 116–117, 122, 136
    fishing industry characteristics   50–54,
        112–113, 115–116, 117–118
    imports   53–54; diagram   55
    landings   31
    management units used by   38, 41, 86–
        87, 107
user rights   177, 183
    allocation   181
    consolidation   182, 183
    informal   178
    transferability   182–183

virtual population analysis   150

warm core rings   11
water circulation   9–11; diagram   10
    effect on eggs and larvae   22
water column   12–13
weight   76–77
wind   6; diagram   17

yield per recruit   144; diagram   149